# COSMIC
# BULLETS

# FRONTIERS OF SCIENCE
Series editor: Paul Davies

# COSMIC BULLETS

## High Energy Particles in Astrophysics

### Roger Clay ▪ Bruce Dawson

#### Foreword by Paul Davies

Helix Books

Addison-Wesley

Reading, Massachusetts

0-201-36083-7

A CIP record for this book is available from the Library of Congress.

Published in Australia by Allen & Unwin Pty Ltd

Addison-Wesley is an imprint of Addison Wesley Longman, Inc.

Jacket design by Suzanne Heiser

123456789-DOH-0201009998
First printing, March 1998

Find Helix Books on the World Wide Web at
http://www.aw.com/gb/

# FOREWORD

In 1992 the fastest object known to mankind hit the Earth's atmosphere 25 kilometres above Utah. When it struck, it was moving at 99.999 999 999 999 999 999 999 per cent of the speed of light, which is the maximum possible speed for ordinary matter. The object concerned was a cosmic ray, or, more accurately, a cosmic particle. Its nature and origin remain a mystery, but it is merely one of myriads of particles that continually rain down upon the Earth from outer space.

Twentieth-century physics is built upon two deep and powerful theories: relativity and quantum mechanics. The former is a theory of space and time, and presents its strange effects most starkly at speeds close to that of light. The latter is a theory of matter, and its effects, which are even weirder than those of relativity, are manifested mostly at the atomic and subatomic level. Cosmic rays combine extreme features of both these fundamental theories of modern physics within a single entity, for they are subatomic particles moving very close indeed to the speed of light. Thus it is here, at the intersection of the two most basic aspects of physical reality known to mankind, that we might expect to see totally new and perhaps bizarre phenomena at work.

Astronomy is probably the most popularised of the sciences. These days most people have heard of black holes, quasars and pulsars. Everyone knows the universe began with a big bang, and pictures from the Hubble Space

Telescope grace our newspapers regularly. Yet cosmic rays are almost unknown outside the scientific community, in spite of the fact that their products penetrate our bodies all the time and they can be a serious cancer hazard to astronauts and even airline passengers.

Elementary particle physics is another glamorous science in its own right. Giant accelerator machines such as Lep (at the CERN Laboratory near Geneva) whirl subatomic fragments around ring-shaped tubes many kilometres in circumference. These technological titans recreate the conditions that prevailed in the universe just after the big bang. They cost billions of dollars to build and run, and require veritable armies of scientists and technicians to operate.

It is often said that however clever an item of human technology may be, Nature will have invented it first. Particle accelerators are a case in point. Somehow, in the depths of space, Nature has created the physical conditions necessary to accelerate subatomic particles to speeds and energies undreamt of at CERN. These particles come to us for free, straight from the heavens. Many of them have travelled thousands of light years across the galaxy, and they bring with them clues about exotic astronomical systems like neutron stars and black holes.

Cosmic rays span the two subjects—astronomy and particle physics—that lie at the cutting edge of physical science. One subject deals with the very large, the other the very small. Historically, cosmic rays were responsible for some of the most important discoveries in subatomic particle physics. In 1932 the first antimatter particle—the positron—was found in the debris from cosmic ray impacts. A few years later the muon and the pion were also found.

During the fifties, sixties and seventies, emphasis shifted away from the study of cosmic rays to man-made particle accelerators for the elucidation of the fundamental structure of matter. However, in recent years the pendulum has begun to swing back. The high cost of particle accelerators is not the only factor. Physicists and astronomers now suspect that Nature is producing exotic types of subatomic particles that are simply beyond the reach of our machines.

Moreover, the study of these particles might reveal crucial details about the objects in space that created them—information that is impossible to glean using telescopes.

Another factor in the renaissance of cosmic ray research is improved technology. The pioneers of this science relied on primitive techniques to obtain their data, such as examining tracks in photographic emulsions or cloud chambers. Today scientists can command a range of sophisticated equipment—high-altitude balloons, ultra-fast electronics and highly efficient detectors.

The authors of this book, Roger Clay and Bruce Dawson, are leading experts in the field of cosmic ray research. In recent years much of Clay's work has been conducted in South Australia with a ground-based system near Adelaide, and in the desert near Woomera in collaboration with John Patterson and several Japanese groups. The system in place at Woomera is designed to seek out high-energy cosmic rays using a method that seems truly astonishing. The experimenters turn mirrors towards the dark desert sky and look for tiny flashes of light created when the products of cosmic rays shower down through the atmosphere. The light has a distinctive signature, analogous to the bow wave of a ship, produced whenever an electrically charged particle travels through a medium (in this case air) faster than the speed of light in that medium.

Miraculous though this technique of detecting cosmic rays may seem, it has proved extremely successful. A much bigger array of detectors, called the Fly's Eye on account of its distinctive shape, operates in Utah, and has been the focus of Dawson's work. This system also scans the sky for light produced by cosmic ray showers, but using an even more sensitive detection method. The results from these observations have been so exciting and intriguing that they have challenged some cherished beliefs in particle physics and astronomy. Particular mystery surrounds the nature of the highest energy particles found recently—particles that by rights couldn't even get here.

Clay and Dawson, who are part of the international collaboration of Fly's Eye users, are now devoting their research to studying these extremely high-energy particles. They are in the forefront of a push to build two really massive cosmic ray observatories, using 16 000 square metres of detectors spread over 3000 square kilometres of terrain, to follow up their preliminary observations. They are convinced that whatever the extremely energetic particles may be, and whatever exotic astronomical systems create them, we will advance our understanding of physics and astronomy enormously by studying them further. As Clay and Dawson point out, the science of cosmic rays is at something of a plateau at the moment. Progress over the last few decades has been impressive, but we can perceive a long climb ahead. The next generation of detectors holds out the promise of making that climb possible.

What will we see from the summit? The whole point of scientific research is to penetrate the unknown. It is tantalising to have a prior glimpse of the view from higher up, but the real joy is in discovering vistas that nobody even suspected. There is a basic property of Nature known to the ancient Greek philosophers as the Principle of Plenitude. It says, roughly speaking, that if something is possible, Nature tends to bring it into existence. With all the resources of the cosmos at its disposal, Nature has produced the entire array of subnuclear entities that we can make in our accelerator machines, and more. Whatever exotic particles remain to be discovered, we can be sure they are out there somewhere.

*Paul Davies*
*The University of Adelaide*

# CONTENTS

# FIGURES

# PREFACE

The Universe as we know it has many strange and enigmatic features. The stars of the night sky have always fascinated us and have been studied for millennia. But they only tell us a small part of the astronomical story.

Recently, we have realised that we need other disciplines such as radio and X-ray astronomy to fill out our vision of the Universe. It is now clear that even these are not enough. It has turned out that the skies around us are full of violence in the shape of extremely energetic sub-atomic particles. The study of the Universe at its most violent is the domain of high energy astrophysics and the high energy particles in question are cosmic rays.

These particles are in many ways deeply mysterious. Despite a century of intensive research, it is still far from certain where they come from or the way they were created. Nevertheless, the little-known story of their discovery and subsequent study has many intriguing twists and wonderful surprises.

It was, after all, thanks to cosmic ray research that physicists discovered the first known particles of anti-matter. Cosmic rays led the way to the discovery of the pion, the glue that holds all atomic nuclei together. They also revealed the presence of a particle called the muon, an entity whose unexpected existence has helped shape modern theories of matter. In short, cosmic rays have offered a natural laboratory for scientists to study particle physics at the highest energies.

Not only that, these particles have revealed new insights into the astrophysical nature of the Universe. Black holes, neutron stars, supernovae and quasars may all be, in some way, implicated in the extraordinary tale revealed by these mysterious emanations from space. This book is our attempt to relate that tale.

Science progresses in fits and starts. It depends on an interplay between observation and theoretical ideas. Sometimes, one has to wait for the other even as routine progress is made. On the other hand, experiment and theory sometimes reach such unexpected conclusions that much time is needed while the practitioners learn to cope with unfamiliar new concepts. The study of high energy particles from the cosmos has had more than its share of unexpectedly bizarre observations and their resulting misinterpretations.

We now seem to be on a plateau in our constant effort to climb the peak of discovery in this field. We feel that we have learned to understand much that would have been unimaginable thirty years ago but we can be sure that there is a long climb ahead and we are presently preparing ourselves for a new thrust into studies of the highest energy particles known to humankind. Already, it is clear that our present ideas of the Universe are not up to the task of understanding what we already know. We need more detail to be quite sure what it is that really needs explanation and what are artefacts of our observational limitations. We will concentrate here on the highest energy material in the Cosmos. Its understanding is our Holy Grail. Experience tells us that unexpected and strange territory in modern astrophysics is still to be reached before the journey is ended.

CHAPTER 1

# THE BEGINNINGS

Even though you neither feel them nor see them, every second as you read this book your body is being bombarded with cosmic rays. Oblivious though we may be to their fleeting presence, we are continually exposed to a hail of cosmic bullets, mysterious particles of matter that appear to owe their origins to the furthest regions of deep space.

To be fair, not all of the high energy particles of radiation that hit us come from space. About half emanate from the terrestrial environment around us. There is radioactive potassium in the salt we eat, uranium in the brick and stone of our homes and radon gas in the air we breathe. In some locations, these sources can be hazardous. Radon gas, for example, is thought to make a significant contribution to the lifetime risk of cancer in regions where its level is unusually high.

The cosmic bullets can be troublesome too. Astronauts and even aircraft crew and passengers are exposed to significantly higher amounts of cosmic radiation than folk on the ground. So much so that at least one airline, Lufthansa, bans pregnant air stewardesses from its flights.

Safety aside, these cosmic bullets have long been the subject of intense scientific interest. What are they and where do they come from? These questions have confounded scientists for over a century. What makes these cosmic rays all the more interesting is their extremely high energies. These entities put human efforts to accelerate sub-atomic particles in the shade. In fact, the lowest energy cosmic rays take off

**Figure 1.1  The energy spectrum of cosmic rays**

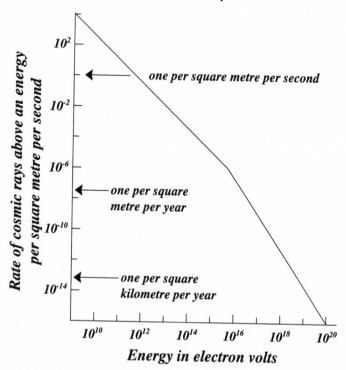

This figure shows how many cosmic rays there are at various energies.

---

where particle accelerators leave off. At their highest energies, cosmic rays are over a hundred million times more energetic than anything humans have ever produced. This begs the big question: how are these super-energetic particles generated?

## The first hints of a problem

Certain aspects of electricity have been known for thousands of years but it is only in recent times that we have learnt to understand them. One of these properties, noted two centuries ago, was perhaps an effect of cosmic radiation. It arose in the study of static electricity.

Most of us are familiar with the fact that if you comb your hair on a dry day, the comb can take on the ability to attract light pieces of paper. On a similar day, you might receive an electric shock when touching the door handle after leaving a car or when touching a metal object after walking on carpet. These effects result from the comb or you receiving a quantity of electric charge through some form of rubbing or friction: the comb in your hair, the car in its motion along the road or your shoes on the carpet. These processes happen because the materials in the comb, car and carpet are made of atoms which themselves consist of a balance of positively charged nuclei and negatively charged electrons. The rubbing process removes some electrons from their atoms. The freed electrons and the remaining positively charged atoms can accumulate so that, eventually, they can have surprising and literally shocking effects.

The most notable early studies of this static electricity were carried out by the French eighteenth-century physicist C.A. de Coulomb which he reported in Paris in 1785. Coulomb's name is most commonly associated with the basic law of electrical charges which describes how strongly charges attract or repel each other. He learned much about electrical charges from a long series of careful experiments as part of his investigations into electrical forces. As a necessary part of this work, he studied the way in which electricity was gradually lost from charged bodies even when, like the balls he used for his tests, they were suspended only by fine insulating strings. In his experimentation, Coulomb found that when he charged anything, the charge did not remain in place forever but was lost in some mysterious way. He reasoned that some of the charge probably got away through the supporting strings but even that did not seem to account for all of the loss.

He found that the speed at which the charge leaked away depended strongly on the amount of moisture in the air. This would not surprise us now since we know that the charging effects of, say, combing our hair, are most noticeable on dry days. This is because on relatively humid days the charge is either conducted away by the moisture in the air or by moisture that has condensed forming a conducting path for the electricity. The question of what other mechanisms there might be for carrying away static electricity was not clarified for another century. In the nineteenth century it was realised that the gradual discharging of bodies could be explained if the air contained freely moving charges. Scientists thought that these could be found on grains of dust charged up when they rubbed against something or were, perhaps, atoms in the air itself.

Today we know that static electricity is produced when electrons are removed from atoms. The atoms then have less than their normal quota of electrons and so carry a net positive charge. In some cases, these positively charged atoms—ions—can move and are likely to be attracted to a region of negative charge where the charges will cancel out. It is now clear that the charge leakage problem that puzzled Coulomb is the result of ionised atoms in the air. But where do those ions come from?

## The discovery of cosmic rays

Near the end of the nineteenth century, the British physicist Charles Wilson carried out an important but, at that time, baffling experiment into static electricity. He measured how quickly charge leaked away from a gold leaf electroscope. This instrument is made of a small brass strip and a narrow piece of thin gold leaf. The gold leaf is very flexible and is attached to the brass by one of its ends so the leaf itself can move freely towards the brass or away from it. When the electroscope is charged up, the charge spreads all over the gold leaf and its brass support because the charges repel each other and spread as much as they can. Being flimsy and light, the whole charged leaf is repelled by the charges in the brass strip and takes up a position at an angle. The size of the angle indicated to Wilson how much

**Figure 1.2  Gold leaf electroscope**

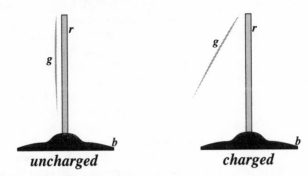

Notes:  g—thin, flexible gold leaf
r —brass supporting rod attached to the gold leaf
b—non-conducting base

When the brass supporting rod and the gold leaf are charged, the charge spreads over them. The charges on the leaf and on the rod repel each other and the flexible gold leaf moves away from the rod. The angle made between the leaf and the rod indicates the quantity of charge.

charge there was. He could measure the rate of charge loss by watching the electroscope's gold leaf slowly descend towards the brass strip.

In an effort to find out what caused the charge to leak away, Wilson put his apparatus in a small sealable container and filled it with dust-free filtered air. He found that the rate at which the charge leaked away was the same regardless of the source of the air. He tried air from the countryside which was probably fairly clean and air from his laboratory which was probably contaminated with city pollution and radioactivity from other experiments. He decided that the leakage could not be due to dust. This was unexpected and quite perplexing. Next, Wilson tried comparing the leakage in the dark and in daylight. Still no difference. Furthermore, there was no change in the discharge rate if the instrument was charged either positively or negatively.

Wilson was forced to conclude that, in some way or another, ions were actually formed within the air in the sealed container at a rate that he could measure with equal amounts of positive and negative charge. Whatever was causing the ions to form did not seem to originate in his apparatus nor in the laboratory. For want of a better name it became known as 'spontaneous' ionisation.

Wilson knew of the ionising effect of X-rays (Roentgen rays as they were called at that time) and of cathode rays (later identified as electrons) as studied at his own Cambridge University by J. J. Thomson. He was also familiar with the effects of radiation from radioactive substances. The 'spontaneous' ionisation had properties very similar to radiation. In 1901, Wilson wondered whether the cause of the ionisation might be radioactive rays from outside the Earth's atmosphere. One way to test this would be to go to a higher altitude to see if the effect increased or to go underground to see if it was reduced. He tried the latter but didn't realise that the discharge effect would be affected by radioactivity in the earth as well as rays or particles penetrating the atmosphere. Unfortunately, his apparatus wasn't sufficiently sensitive to separate the effects. The measurements he made at night in a Scottish railway tunnel showed no measurable reduction in ionisation effect and seemed to eliminate the possibility of a cosmic origin. Baffled, he concluded that the source of the ionisation must be something in the air itself.

This issue was not clarified for another decade. During this time an improved version of the simple electroscope used by Wilson was developed. These new 'ionisation chambers' were more robust and more accurate than the electroscope and became standard equipment in radiation laboratories. One was used by the Curies in their pioneering studies of radioactivity. It was an ionisation chamber that finally helped solve the problem of the charge leakage.

The crucial experiment took place on 7 August 1912 and resulted in a Nobel Prize though not, alas, for Wilson. The experiment involved ascending in the atmosphere, as opposed to making measurements at depth as Wilson had done. The scientist was Viktor F. Hess who, in a heroic series of flights, carried ionisation chambers to great altitudes in the gondola of a hydrogen balloon. Cosmic ray experiments have had a tradition of being carried out in inhospitable places but few have matched Hess's intrepid adventures. In these, he rode in a small basket under a large volume of highly flammable hydrogen gas to an altitude above 5000 metres. Each measurement of the ionisation chamber took an hour and Hess made measurements during both the ascent and descent of the balloon. The flights were all the more hazardous since Hess, unlike modern mountaineers, did not have the advantage of being able to carry extra oxygen. At such heights his judgement could well have been affected by the thinner air.

Often scientific breakthroughs result from the availability of a new instrument. In the case of the balloon observations, such an instrument was the much improved ionisation chamber. It was designed specifically for the balloon experiments by a Jesuit priest, Father Th. Wulf. In 1909 Wulf had himself made measurements at the top of the Eiffel Tower in an unsuccessful search for a change in the amount of ionisation as the distance from the surface of the Earth was increased. The Wulf detectors were compact and rugged and were designed to cope with the great changes in atmospheric pressure as a balloon rapidly rose. Hess carried three such radiation detectors on his flights.

Hess was rewarded by finding that as his balloon began its ascent the ionisation (or the radiation causing it) fell off a little. However, above an altitude of about 2000 metres the level of radiation slowly increased followed by a clear and rapid increase as the highest altitude of 5350 metres was approached. His interpretation was that the radiation from the radioactivity of the earth had an influence up to about 2000 metres, but at higher altitudes something else caused the ionisation to increase. Hess guessed that this other component was a 'radiation of high penetrating power [which] enters our atmosphere from above'. On one occasion he took a flight during a solar eclipse. As there was no reduction in the ionisation effect nor was there a reduction at night, the source of the radiation was unlikely to be the Sun.

By this time, natural radioactivity had already been extensively studied, although the nuclear processes causing it were not at all understood. Lord Rutherford, working in Montreal, had established that the radiation fell into three broad categories. These were alpha particles, now known to be high-speed helium nuclei; beta particles, electrons which are also known as cathode rays when they are man-made; and gamma-rays which are energetic 'photons' with the same basic properties as light or X-rays, but with more energy. These forms of radiation have quite different properties. The alpha particles only travel tiny distances, beta particles are more penetrating and gamma-rays are by far the most penetrating. Hess's immediate reaction was to think that his ionising component consisted of gamma-rays, because if the radiation came from space it would have to penetrate many kilometres of atmosphere. The only radiation then known to be remotely capable of such a feat was the gamma-rays. As we shall see, this idea turned out to be incorrect though neither Hess nor anyone at that time could have realised why.

Hess's adventures were followed by some even more daring balloon flights undertaken by the physicist Werner Kolhorster. As a result of studying radioactivity, he continued to develop the Wulf ionisation chamber and, as a young man in his mid-twenties, made a series of balloon ascents in 1913 and 1914. He reached an altitude of 9300 metres, similar to the altitude of Mt Everest or the cruising altitude of passenger jets, where he measured the ionisation effects of the radiation to be fifty times its value on the ground. In 1930, Kolhorster established in Potsdam the first institute specifically for cosmic ray research.

Hess's and Kolhorster's observations of increased levels of radiation at high altitudes clearly suggested that the radiation came from a source somewhere outside the environment of the Earth. But from where? At this time, astronomy was barely equipped to begin to address this problem.

The way that the radiation made its way through the atmosphere was also unclear. It's one thing to notice that there is more radiation higher up, but it is much more difficult to work out how the radiation intensity should vary. Early ideas, which had cosmic gamma-rays simply reducing in intensity at greater and greater depths in the atmosphere turned out to be quite wrong. We now realise that this error was inevitable since most of the particles at higher altitudes are not gamma-rays at all. The way forward proved to depend on several initiatives. One was to study how the intensity varied across the Earth according to latitude, pressure and other parameters. As we will see, another area of progress was to be in the development of new instruments that enabled researchers to observe individual particles of radiation.

# CHAPTER 2

# NEW TECHNIQUES TO UNDERSTAND COSMIC RAYS

In the early years of the twentieth century, research was very labour intensive. Experimentalists were obliged to record most results by hand, perhaps from readings of meter dials or from photographic film. Today such data gathering and analysis has, of course, been transformed by the use of electronics and computers. In cosmic ray research this transformation has gone hand in hand with the development of the new science of nuclear physics.

When Wilson, Hess, Kolhorster and their colleagues conducted their experiments, the main measuring instruments were the gold leaf electroscope and its later incarnation, the Wulf ionisation chamber. With these, they established that the air around us is continually being ionised at a low level which explains why charge could leak away from an electrified object.

But to quantify the effect more precisely required new tools. One of the most important was devised by the German physicist Hans Geiger who worked in Lord Rutherford's laboratory in Manchester. In 1908 he developed a device which clicked each time a particle passed through it. The Geiger counter is now synonymous in the popular mind with the measurement of radiation. Meanwhile, Wilson, who had done so much to get this subject moving, was working on another experimental apparatus that also turned out to be of immense importance to the study of particle physics. He developed an instrument that could actually register the paths of individual ionising particles.

This was the famous cloud chamber, which was followed some years later by the bubble chamber. Those instruments, together with electronic counting techniques, revolutionised the study of particle physics for the next half-century.

## The 1920s—consolidation

In the 1920s interest in cosmic radiation studies increased rapidly in response to the realisation that the ionisation of the air was the key to new and fundamental ideas. Radiation, so recently discovered coming from disintegrating nuclei, now also seemed to be coming from space. This new radiation was studied throughout Europe and interest was also strong in North America. Some fundamental results were soon recognised but not so quickly understood. Extensions of the balloon observations showed that, surprisingly, there was an altitude at which the radiation was greatest and this was *not* at the top of the atmosphere. This did not seem to make sense if the radiation was from beyond the Earth. One would expect the atmosphere to progressively absorb the radiation, not build it up and then absorb it.

For a time, it was thought that there must be a source of radioactivity high in the atmosphere but it then became clear that the radiation was not the same as that from radioactivity since it could be much more penetrating. It was found at great depths below the surfaces of lakes and glaciers. Measurements in tunnels also showed that it could be found under considerable depths of rock although its intensity was greatly reduced. Even the most penetrating gamma-rays from natural terrestrial radioactivity had nothing like this penetrating ability. By 1925, it was clear to most that the origin of the phenomenon was indeed extra-terrestrial and the influential American physicist R.A. Millikan chose to name it 'cosmic rays'. Millikan, who is most famous for his oil drop experiments in which he measured the charge of the electron, even then believed cosmic rays to be very energetic gamma-rays, then the most penetrating particles known.

11

That view became untenable almost immediately. Gamma-rays, like light, travel in straight lines. If cosmic rays were just gamma-rays and if they were produced in our Milky Way galaxy, we would have a region in the sky with strong cosmic ray intensity just as we have a bright Milky Way observable with light. In 1926, measurements were made in South America which showed that the intensity of cosmic radiation was essentially no different when the Milky Way was overhead and when it was not.

The clue which led to the next step forward in understanding came from careful measurements of how the intensity of cosmic ray ionisation varied over the Earth. This small variation is not due to a different view of the stars in the sky but it turned out to relate to the magnetic field of the Earth. This magnetic field cannot affect gamma-rays but it does affect charged particles. At about this time it had been realised that the Aurorae, beautiful coloured lights sometimes seen in the sky near the North and South Poles, are due to high energy particles from the Sun penetrating into the atmosphere. These particles, being charged, are diverted by the Earth's magnetic field into complex paths that eventually hit our atmosphere, making it glow.

Aurorae are seen mainly close to the (magnetic) poles of the Earth because that is where the lines of the Earth's magnetic field curve down into the Earth. The passage of charged particles is difficult *across* the direction of the field lines and the cosmic rays tend to follow them. So, few particles reach the Earth near the equator where they have to cross many horizontal magnetic field lines. Near the poles, the lines are directed almost vertically downwards and the particles can easily follow the lines inwards to the atmosphere where they make its molecules glow producing the Aurorae. If the particles are particularly energetic, they are channelled less by the lines and the Aurorae are visible further from the poles. In this way, a knowledge of the Earth's magnetic field plus observations of Aurorae give us an almost direct measure of the energies of the auroral particles from the Sun. In the same way, it was recognised that the variation of cosmic ray intensity with distance from the pole should give information on the energies of the cosmic ray particles.

Measurement of this latitude effect was not easy. The ionisation chambers in use in the 1920s had become quite sensitive but the experiment entailed transporting chambers over large distances by sea. It was a challenge to keep them accurately calibrated while the measurements were made. The results of the observations were first reported in the late 1920s by the Dutch scientist, J. Clay, following his travels to the East Indies. More sophisticated measurements continued into the early 1930s. At sea level, there is a drop in intensity from mid-latitudes to the equator of about 6 per cent and at higher altitudes the effect is considerably greater. There was now no doubt that the cosmic rays included a charged particle component from outside the Earth's magnetic field but the surprise was the size of the effect. It could be consistent with *all* of the radiation being in the form of charged particles, not just a small addition to a gamma-ray component.

A further check can be made on the relative numbers of charged particles and gamma-rays in cosmic rays. Because we know the polarity of the Earth's magnetic field (a 'South' pole at the north and a 'North' pole at the south), we can predict how incoming charged particles will be deflected by it. This in turn will depend on whether the particles are positively or negatively charged. The test is to look to see whether more cosmic rays come from the east or from the west (when observations are made roughly at the equator). This East–West effect was predicted by Georges Lemaître (probably best remembered now for his work in cosmology) in Belgium and Manuel Vallarta in Mexico and independently by Bruno Rossi in Italy. Rossi decided to look for the effect and, realising that it would be most pronounced at high altitude and near the equator, prepared an experiment which was carried out on a mountain in East Africa. The East–West effect was found but the Italian experimenters were painfully disappointed to discover that they had been beaten by a few months with

13

**Figure 2.1 The coincidence technique**

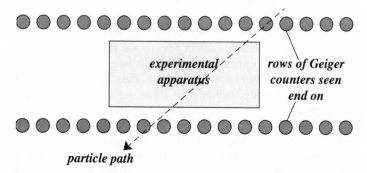

*experimental apparatus*

*rows of Geiger counters seen end on*

*particle path*

By selecting which pairs of counters are traversed by a cosmic ray, it is possible to select its arrival direction. It is also possible to ensure that any recordings of data from the experimental apparatus (such as a cloud chamber) are only made when a cosmic ray actually passes through it.

an observation made in Mexico City. Not only that, but only the Lemaître and Vallarta prediction was credited. There was no mention of Rossi's work. To Rossi it all seemed very frustrating.

Observation of the East–West effect required the use of Geiger–Müller counter tubes, an improvement on the Geiger counters mentioned earlier. Some cosmic rays can pass through several of these tubes without stopping. The tubes can then be arranged in groups which respond only when certain combinations register the passage of a single particle. If you specify that only tubes in a certain straight line must be hit, then the direction of the particle is specified by the orientation of the counters. The tubes are said to be operated 'in coincidence' and will count most quickly if they are oriented so that the direction of the incoming particles is vertical. That is the most direct path through the atmosphere and the incoming particles then suffer least absorption. One might expect that the counting rate would reduce progressively with angle from the vertical and so it does. On the other hand, the rate changes differently with angle from the vertical depending on the geographical orientation of the counters. There is a clear East–West effect (in equatorial regions) with more particles coming from the west than the east. The effect itself was expected but the polarity of the results was a surprise. The predominance of particles from the west showed that the particles were not negatively charged electrons as had been assumed but were positively charged. This result turned out to be of fundamental importance in eliminating many misconceptions but its true meaning was not recognised until more pieces of the puzzle fell into place.

## Particles and cloud chambers

The first three decades of the twentieth century were amazingly heady times in physics. Just before the turn of the century, some scientists had come to the conclusion that there was little more to do in physics than fill in a few more figures after the decimal point of various fundamental constants. They couldn't have been more wrong. The edifice of classical physics began to fall apart with the coming of the new century. The first signs of the revolution in atomic physics emerged with the work of the German physicist Max Planck. He developed a mathematical technique for explaining the spectra of heated bodies. A spectrum is the detailed distribution of colours in light such as we see in a rainbow. Hot bodies may become 'red hot' and even hotter ones are 'white hot'. The change in colour is our way of interpreting a change in the spectrum of the light from a heated body as its temperature increases. Planck's mathematical theory assumed that light was produced in discrete lumps (later called quanta) with quite specific energies. This was a new idea because physicists had previously assumed the properties of light to be continuous. Almost immediately, Albert Einstein showed that these quanta were more than mere mathematical conveniences and that light really does exist in the form of packets of energy. These became known as photons.

This 'quantum' description of light applies to all forms of electromagnetic radiation from radio waves at one end of the spectrum, to visible light in the middle and gamma-rays at the far end. In everyday life, when we use light or radio waves, we are using photons with individual energies so small that we usually do not have to consider that the total energy is made up of lots of small lumps of energy. So, for practical purposes, we can usually forget that they exist. For instance, a light globe emits billions of billions of photons per second and we seem to have a continuous flow of energy. Intriguingly, the idea that light came in particles had been advocated in the seventeenth century by Isaac Newton. This so-called 'corpuscular' theory was seemingly disproved by experiments in the eighteenth century which convincingly demonstrated that light was a wave phenomenon. In the twentieth century, physicists found that light had certain properties which required it to be regarded sometimes as a collection of particles and sometimes as a continuous wave. This surprising state of affairs was further embellished when it was discovered that electrons, which were naturally assumed to be particles or lumps of matter, also exhibited wave properties. There is no doubt that this period saw one of the greatest upheavals in the physicist's view of the world.

Another ingredient in this revolution was a fundamental revision of the concepts of space and time. We all have an intuitive feeling of the meaning of space and our place with respect to other bodies on Earth or in the Universe. We are comfortable in the feeling that our position relative to other objects is fixed and can be uniquely defined. We also have an intuitive feeling about the concept of time and its inexorable uniform onward march. These comfortable concepts were abandoned by physicists as they absorbed the impact of Einstein's Special Theory of Relativity which was also derived at about the turn of the century (see Appendix 1).

This theory is fundamental and is, for many physicists, an everyday tool to be used for calculations about the ways in which bodies move and interact. It gives the intuitive answers we are used to when we deal with many everyday circumstances but also gives the correct answers under more extreme conditions. The Special Theory, at least, follows from much of nineteenth-century physics. Its equations were known earlier but, like quantum ideas, were not recognised to be fundamental to physics. Einstein showed that the equations were not just mathematical conveniences but that space and time really were inextricably linked into a fourfold combination of three-dimensional space and a dimension of time. Amazingly, these space and time dimensions could be exchanged, time for space and vice versa. Einstein also showed that in order to retain certain sacrosanct principles of physics—namely the laws of conserving energy and momentum—some intuitive ideas had to be abandoned. Mass was not fixed but could be exchanged for energy. Matter could be made from energy and energy from matter. The old idea of energy had to be extended to include a component of energy associated with mass. The relationship between mass and energy turned out to be crucial to understanding the observed properties of cosmic radiation.

We return to Charles Wilson and his first love in physics, the clouds. Wilson spent several weeks in 1894 at an observatory on the summit of Ben Nevis, the highest point in the British Isles. He was fascinated by the cloud phenomena he saw there and wanted to study such phenomena under laboratory conditions. His approach, well known before this experiment, was to place moist air in a test chamber and force it to expand. This process causes the air to become supersaturated with water vapour which condenses to form a fine mist in the same way that clouds form in the sky. Condensation usually requires the presence of tiny particles or 'nuclei' to seed the formation of water droplets. Air naturally contains dust particles which perform this service.

Wilson found that if his laboratory-made 'clouds' were allowed to settle to the bottom of the chamber, they carried the dust particles with them. So, after a few cloud-forming expansions the dust was completely removed and clouds would not form under modest expansion of the air. It is not so easy to study this phenomenon in the laboratory because as the expansion takes place the air cools (just like the gas from an aerosol can cool your finger) and then warms again as heat enters from the outside. Wilson constructed a special apparatus to make the expansion very quick so that he didn't have to allow for the slow influx of heat. After the dust had been removed, no cloud at all could be seen until the air was expanded by a factor of at least 1.252. At larger expansions still (up to 1.375), Wilson saw that a small shower of drops, like rain, was produced and at even greater expansions dense clouds formed even without the dust. This was the phenomenon which originally interested Wilson while on Ben Nevis. Wilson became intrigued by the intermediate stage in which there was a steady shower of drops. He recognised that this must mean that there was a steady production of small nuclei to initiate the formation process.

The year 1895 was a crucial one for physics. In November of that year the German physicist Wilhelm Roentgen discovered X-rays. His observations were to be repeated worldwide within a few short months. Meanwhile, Wilson was working at Cambridge in the Cavendish Laboratory under the great experimental physicist J. J. Thomson, who was himself experimenting with X-rays as early as the beginning of 1896. Wilson was allowed to use Thomson's X-ray apparatus to see what effect they might have on his supersaturated vapours. He was delighted to discover that the X-rays (which can cause ionisation) immediately produced a dense fog in his chamber when the expansion was such as to previously produce only a steady shower. The 'rain' had previously been revealing the presence of natural ionisation in the chamber, some of which was due to the then unknown cosmic rays. The charged ionized atoms had replaced dust particles as the nuclei which allowed condensation to form.

Wilson's interest moved to measurements of the conductivity of the air but he returned to the production of droplets in about 1910. At that time, the nature of radioactivity had become clearer with the contributions of

Rutherford and his co-workers. It had been realised that alpha- and beta-rays consisted of charged particles and Wilson asked himself whether their paths might be traced out by cloud condensation on the ionised atoms which they produced as they passed through a gas. He spent a great deal of time trying to design the optimum apparatus for producing, illuminating and photographing the tracks. Before the design was properly complete, he decided to try anyway and was rewarded to see wispy tracks when he shone X-rays into the chamber. He also looked at tracks from alpha-particles and showed their photographs to W.H. Bragg, then the expert on alpha-particles. Bragg had just published a diagram showing the type of paths he expected from alpha-particles, deduced by indirect means. The diagram and the real photographs were strikingly similar.

The cloud chamber technique was used with some success through the 1920s but came into its own at the end of that decade. We have seen that cosmic radiation was very penetrating, much more so than natural radioactivity. It was important to ask why that might be so. It could be that the radiation was some completely new phenomenon or, perhaps, it was just a more energetic form of the same particles which were then becoming familiar: the electrons, alpha-particles and gamma-rays. A way of testing this would be to view the tracks of the cosmic rays as they passed through a cloud chamber while in the presence of a strong magnetic field. Charged particles are deflected in a magnetic field by an amount which depends on the strength of the field, on the magnitude and polarity of their charge, and on their momentum (or energy). The main unknown factor was the energy and this could be estimated from the deflection of the track. Carl Anderson and Robert Millikan in the United States built such an apparatus in 1929.

The technique was perfected over the next decade by Patrick Blackett and his co-workers in England. The crucial development was to use a combination of the Geiger counter and the cloud chamber. Cloud chambers were at first set to record at random times and it was hit and miss whether an interesting track would be present. However, when their recording was initiated by the detection of a particle in a Geiger counter above the chamber, pretty well every photograph contained something of interest. The cloud chamber quickly showed electrons with energies up to a billion electron volts, a thousand times greater than previously known from radioactivity. These huge energies were an important clue to their penetrating power.

This observation alone would have justified the effort put into the project but the discovery of most far-reaching significance was made in August 1932 when Carl Anderson, a colleague of Millikan at the California Institute of Technology, realised (as did Patrick Blackett and Giuseppe Occhialini in England at almost the same time) that some of the particle tracks he photographed were due to positrons, the identical but positively charged twin of the electron familiarly found in atoms. Such 'anti'- particles had been predicted by the great British physicist Paul A.M. Dirac not long before although, at the time, Dirac had thought that the proton was the positive particle in his prediction. Anderson found that many of the positrons resulted from the interaction of cosmic rays with the nuclei of atoms. However, about six months later he realised that sometimes, when a photon (a gamma-ray) impinges on a nucleus, a pair of particles is formed consisting of both an electron and a positron. The energy of the gamma-ray is clearly converted directly into the masses and kinetic energies of the secondary particles just as had been predicted by Einstein's Special Theory of Relativity. The discovery of the positron and the confirmation of the existence of anti-particles was one of the most significant experimental results in the history of physics. The process of converting the energy of a gamma-ray into a particle and its anti-particle is now called 'pair production'.

The early 1930s were perhaps the heyday of the cloud chamber as a tool for investigating the properties of the so-called elementary particles, although the technique is so versatile that it is still occasionally used to the present day. A number of experimenters worked creatively with cloud

**Figure 2.2  Cloud chamber photographs of cosmic ray events**

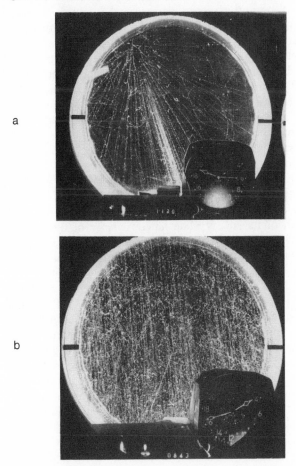

a

b

These photographs show the tracks of secondary cosmic ray particles.

      a—A shower begun by an interaction in the wall of the chamber.

      b—The core of a large cascade.

*Photos*:  B.D. O'Donnell.  _____

21

chambers and many important results were found. We need to record here that some particle pairs appeared to be produced by electrons. It was recognised that, in this case, the electron first produced a photon (gamma-ray) as it passed close to a nucleus and the gamma-ray then produced a pair. The process of an electron emitting a gamma-ray is called bremsstrahlung, translated as 'braking radiation', since it allows the collision of an electron and a nucleus to cause a slowing of the electron and still conserve energy and momentum.

Notice that we now have a new possibility. An electron can produce a gamma-ray as it loses energy and the gamma-ray in turn can produce an electron and a positron. One particle (the electron) has become three (two electrons and one positron), although the original electron will have lost much of its energy. The kinetic energy of the original electron is converted into mass and kinetic energy to be shared with the other particles. An electron is the particle of lowest mass. The mass of this particle in energy units is about 500 000 electron volts, usually written as 0.5 MeV (1 MeV is one million electron volts of energy). This value is quite small compared with the kinetic energies Anderson found for many cosmic ray electrons (many thousands of MeV). As a result, the energy of a single cosmic ray electron can be converted into the masses of many secondary particles. This repeated process of bremsstrahlung and pair production to progressively convert energy into more and more secondary particles is called cascading. Since this particular type of cascade involves only electrons (positrons are often regarded as the same particle but with an opposite charge) and gamma-rays, it is often called an electromagnetic cascade or shower.

## Early studies of cosmic ray showers (extensive air showers)

Showers produced in the Earth's atmosphere by cosmic rays were studied through the 1930s after the first evidence for their existence was found in 1927–29 by Pierre Auger

22

and Dimitry Skobeltzyn while Skobeltzyn was visiting Paris from Russia. The showers were clearly energetic but they appeared to contain a component other than the now familiar electrons, positrons and gamma-rays. This further component was more penetrating and it was these secondary particles which were measured in tunnels or at great depths in lakes, often as single isolated tracks in cloud chambers. Such particles had to be something different and new. For want of a better term they were labelled x-particles.

In 1936 it was confirmed by Millikan's cosmic ray group that there were indeed two major distinct groups of singly charged particles reaching the ground: the electrons and the x-particles. Anderson from that group was awarded the Nobel prize with Hess for his cosmic ray researches that year and gave an address in Stockholm. He commented that the highly penetrating particles 'although not free positive and negative electrons' were likely to be interesting material for further study. It was to be so.

In 1935, a quite separate development led to a mistaken idea about the identity of these particles. The Japanese theorist Hideki Yukawa proposed that a new kind of particle was needed to explain the nature of the nuclear force that holds together atomic nuclei, now known as the strong force. Until then no-one had any workable theory of the nuclear force. Yukawa's idea required a particle of mass intermediate between electrons and protons. Just two years later, the x-particles were shown to have a mass of about 200 electron masses in cosmic ray cloud chamber experiments by Seth Neddermeyer and Anderson. It was widely assumed that this particle was the very one Yukawa had predicted. Being intermediate in mass between the electron and the proton, it was dubbed the mesotron (from the Greek 'meso' for middle) though this was soon shortened to meson, a name vehemently hated by Millikan. However, as we shall see, this conclusion that the mesons in cosmic rays were also involved in the strong nuclear force turned out to be incorrect—the required particle was the pion which was not discovered until some years later. As time went by, the cloud chamber particles were renamed mumesons and are now known simply as muons. They are not even classified as mesons today.

Strangely, the muons seemed to be absorbed in the atmosphere more rapidly than expected from laboratory measurements of their absorption. It was suggested that this might be because some of the muons decayed in their long passage through the air and this was confirmed in 1939. When the lifetime of a muon is measured in the laboratory (as was first done by Rossi in 1940) it is found to be so short (2 microseconds) that at close to the speed of light, one could reasonably only expect a muon to travel less than a kilometre. But they easily pass through many kilometres of air to reach us. Rossi recognised this as an example of time dilation in Einstein's relativity. Extraordinarily, given that Special Relativity had been proposed in 1905, this was the first time the time dilation effect had been experimentally confirmed. The fast-moving muons have their apparent lifetime extended simply because their speed is close to that of light.

An important question then was what relationship these muons had with the electromagnetic cascades. The intensity of the muons seen in cloud chambers did not seem to depend on altitude like the electromagnetic particles. This suggested that their origins were not closely tied. Pierre Auger and his co-workers in France felt that the bulk of the observed electromagnetic particles could be just the final stages of electromagnetic cascades initiated by primary particles (presumed to be electrons) high in the atmosphere. Their picture was that the cascade of elementary particles would travel more or less together through the atmosphere because all of the particles would have very high energies compared to their rest masses and so would all travel at close to the same speed—the speed of light. On the other hand, at each interaction with an air atom or electron, the particles would be pushed to the side a little in a random way (called Coulomb scattering since it is due to the electric fields of the charged particles) and, as a result, the shower would extend sideways making it into the shape of a thin saucer. This structure led to the phenomenon being named an Extensive Air Shower (or EAS). At first, the name Auger Shower was used.

Auger and his colleagues had seen showers in cloud chambers but even a large cloud chamber had a size which was much less than a metre across and this was impossible to extend to any great extent. However, electronic counters such as Geiger counters could be spaced horizontally and arranged to operate 'in coincidence' so that an electronic circuit recognised when two or more had detected shower particles at identical times. These particles must have come from the same shower. Such experiments had already been carried out for horizontal counter spacings of up to half a metre and it had been seen that as the counters were gradually separated the rate of coincidences quickly reduced. The effect was probably first seen by Bruno Rossi in 1933 on his expedition to East Africa to observe the East–West effect but, as he reported, he had 'lacked time to study more closely this phenomenon'. Auger found that his counters continued to register coincidences up to 300 metres apart! The rate of coincidences was now reduced by 1000 times but still showed no evidence of stopping altogether.

It looked as though showers were initiated high in our atmosphere by cosmic ray particles from space. The showers which were observed at the ground with Geiger counters and cloud chambers contained electrons, positrons and muons. These particles were presumably produced in interactions between the cosmic ray particles and atmospheric gas atoms. The muons were created in the early primary cosmic ray interactions but the electrons and positrons in the shower were progressively produced in electromagnetic cascades in the air. However, shower theory suggested that electromagnetic cascades should not spread out by more than a few tens of metres and so there had to be another explanation for the large lateral spread of the observed showers. The answer appeared to be the penetrating muons. These were the only particles capable of passing through large thicknesses of rock and water. They also travelled easily and almost in straight lines through the air. Thus

**Figure 2.3  A cosmic ray extensive air shower**

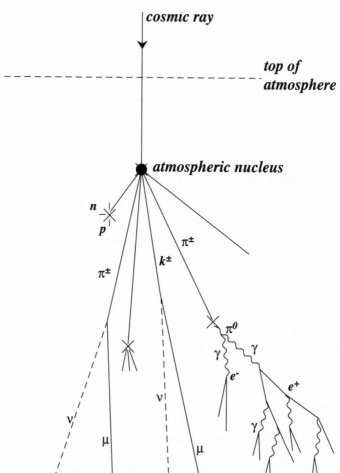

The major components of a cosmic ray extensive air shower.
Pions ($\pi$) and kaons (k) are produced in the initial interaction
along with protons (p) and neutrons (n). Charged pions may
decay to produce muons ($\mu$) and neutrinos ($\upsilon$). The
electromagnetic cascade of electrons (e-), positrons (e+) and
gamma-rays ($\gamma$) is initiated by the decay of a neutral pion.

muons, if they were produced high in the atmosphere travelling at even small angles to the direction of the main part of the cascade, would reach the ground at large distances from the central regions of the shower. For instance, a muon setting out from an altitude of about 10 kilometres at an angle of only one degree to the shower direction would be over 150 metres from the shower centre (usually known as the core) on reaching the ground. Our picture of a cosmic ray shower is now a combination of a central energetic core plus electromagnetic and muon components. The electromagnetic component dominates within a few tens of metres of the shower and the muons are the most important component further out.

An astonishing result of the work done by Auger was his calculation of the energies associated with his showers. Each of the particles in a shower was known from theory and confirmed by cloud chamber measurements to have an average energy of tens of millions of electron volts. Auger was able to estimate the total number of particles in one of his large showers since he knew from his coincidence measurements how the number of particles varied with distance from the centre of the shower. These large showers proved to have a million or more particles within them making a total shower energy of at least 1 million (the number of particles) times 10 million (the average energy of each particle) electron volts. This calculation makes no allowance for energy being lost by the shower in its passage through the atmosphere and its value is now known to be low by ten to one hundred times. Auger had thus measured energies coming from the Cosmos which were at that time inconceivably high and it was critical that some explanation for them was found. As a comparison, the energy of photons used in optical astronomy is about 1 electron volt or about one thousand million million times less than the cosmic ray. We now know of a number of ways in which particles *could* reach these energies in the astronomical objects we'll describe later. Many of them certainly operate but, to the present time, we have been unable to locate one single dominating explanation for how single particles could be accelerated by Nature to the highest energies known.

## Particle physics and cosmic rays

In order to interpret the cosmic ray phenomena known at the start of World War II, only a few particle types appeared necessary. There were the nucleons (protons and neutrons), electrons (including positrons, their anti-particles), gamma-rays, neutrinos and muons. Theory also needed mesons as the particles which were involved in the strong force which held nuclei together against the repulsion of the positive charges on the protons. This meant that the mesons would have to interact strongly since the nuclear glue which they provided needed to be very strong. They would also interact strongly with air nuclei if they were liberated from their own nucleus in a cosmic ray interaction. As a consequence, their lifetime outside the environment of the nucleus should be very short. Unfortunately, the muons (which were thought to be these mesons) observed in cosmic ray showers clearly had a long lifetime—sufficient for them to reach the ground before decaying when allowance was made for their extended lifetime due to relativistic effects. The mesons responsible for the strong force had to be different to those muons detected in showers.

The search for an understanding of the properties of the muons and other high energy particles provided by cosmic rays continued with a growing portfolio of obser-vational techniques. Cloud chambers were powerful research tools due to their high efficiency when their operation was triggered by other detectors. However, for a time, the use of photographic emulsions became the technique which provided discoveries in particle physics. This method was first devised by Cecil Powell at the University of Bristol in England just after World War II. The technique works because when high energy particles pass through a photo-graphic emulsion (such as the film in a camera) they leave a record of their passage which can be made visible by developing the film.

**Figure 2.4 Family of sub-atomic particles known in the late 1940s**

| Name | Comment | Rest mass | Lifetime |
|---|---|---|---|
| proton ⎫ | nucleons, i.e. particles making up the bulk of a | 1 GeV | infinite |
| neutron ⎬ | nucleus | 1 GeV | 15 minutes |
| electron ⎫ | anti-particles (leptons) | 0.5 MeV | infinite |
| positron ⎬ | | 0.5 MeV | infinite |
| gamma-ray | photon | 0 | infinite |
| muon | A massive lepton like an electron. Now known not to be one of the family of mesons although it was first named the mu-meson | 106 MeV | 2.2 micro-seconds |
| pion | A meson. Part of the family of strongly interacting particles known as hadrons | 140 MeV | 26 nano-seconds (charged) $0.8 \times 10^{-16}$ s (neutral) |

A related technique is currently used for monitoring the exposure of radiation workers who carry a small film badge which can accumulate a record of the radiation exposure of the wearer. Photographic film contains quite heavy nuclei which make effective targets for particle interactions. Also, the films themselves require little or no ancillary apparatus when in use, although their development and scanning by microscopes became a specialist art when the technique was used for studying particle physics. The films are produced with thick emulsions so that as complete a picture of the cosmic ray interaction as possible is recorded. The cosmic ray exposure takes place when the films are taken to high altitudes where there are particles of great energy still in the cosmic ray cascade. (Some are even the primary particles themselves.) The location could be simply a high altitude ground site or, possibly, the payload of a balloon or a rocket.

In the immediate post-war period, the University of Bristol in England was the Mecca for such work. A great success occurred when a second type of 'meson' was found in 1947 using this technique.

This new particle, to be named the pion, was produced in a highly energetic interaction and quickly decayed into a second particle, the previously known muon. This true Yukawa particle was found to come in three varieties, positive, negative and neutral. The positively and negatively charged pions decayed to give the positively and negatively charged muons in showers. The neutral pions decayed almost instantaneously into pairs of gamma-rays. These initiate the electromagnetic cascades in showers. All seemed well. The collection of particles necessary to explain the observed properties of cosmic rays was complete.

We now have a picture of the cosmic rays from space being mainly positively charged nuclei (protons, helium nuclei etc.) with a huge range of energies. When one of these hits the atmosphere, its interactions result in the production of pions. The charged pions decay in a short time to produce energetic muons which are able to pass through the atmosphere and reach the ground. The neutral pions decay almost instantaneously into gamma-rays which begin electromagnetic cascades. Although each individual cascade is relatively short-lived, the original particle continues to travel deeper into the atmosphere, progressively losing energy and causing further electromagnetic cascades to begin. As a result, the particles at ground level are a mixture of the muons and the electrons, positrons and gamma-rays from the electromagnetic cascades. There can also be a central energetic core from the remnants of the original cosmic ray.

# CHAPTER 3

# COSMIC RAYS AND THEIR UNIVERSE

The 'cosmic ray' particles we encounter at the surface of the Earth are not the original particles which have perhaps travelled from distant galaxies. They are 'secondary' cosmic rays created out of the energy of the primary particle. These secondary particles appear in complex particle cascades, which penetrate the atmosphere. Such cascades, which take the form of showers of particles, reveal much about the 'primary' cosmic rays. The highest energy primary cosmic ray particles known have energies of 50 joules in a single particle which corresponds to the power from a light globe in about one second: not a lot by our measures but absolutely enormous for a single particle. Unfortunately, the arrival rate of such energetic cosmic rays is miniscule. The rate is typically measured in particles per square kilometre per century. However, lower energy cosmic rays are much more numerous and at the lowest energies can be quite intense. Something like a hundred low energy cosmic particles pass through our bodies every second when we are at ground level.

Cosmic ray studies have become an important part of astrophysics. Although their origin is far from certain, it is thought that they tell us about processes in the most exotic environments of the Universe: radio galaxies, quasars and the seething rotating disks of infalling matter around neutron stars and black holes. Our understanding of these astrophysical objects is crude and a major thrust of current cosmic ray research is to understand how Nature might produce such extraordinarily energetic particles in these objects.

A fundamental stumbling block of these studies is that the primary cosmic radiation consists of charged particles. Unlike the particles of light used in some form or another by almost all other fields of astronomy, these charged particles are swerved by the magnetic fields which we now know to permeate the whole of the Cosmos. The result is that in almost all cases charged cosmic ray particles carry no record of their direction of origin and we cannot perform positional astronomy. Measurements of their arrival directions carry useful information on the cosmic magnetic fields but do little to help identify the cosmic ray origins.

Nevertheless, we have observed the cosmic ray intensity to increase and a flare to occur on the Sun at the same time so we know that some cosmic rays, at their lowest energies, come from dramatically explosive occurrences on our Sun. At higher energies, exploding stars called supernovae or their remnants are probably responsible for the acceleration of particles since radio astronomers observe radio waves produced by energetic cosmic rays in such objects. At the highest energies, we have to look at the most extraordinary objects in the Universe to speculate on the cosmic ray origins. Cosmic ray astrophysics therefore encompasses many of the most spectacular and unusual members of the astronomical 'zoo'. To understand the significance of cosmic rays we need to explore the range of properties of this zoo. It is a story in which cataclysm plays a big role. Active galaxies, supernovae and neutron stars all loom large but before we plunge into their murky depths let us look at the modern picture of the whole Universe.

## Our place in the Universe

Our Sun sits out towards one edge of an enormous spiral galaxy, the Milky Way, a pancake-shaped collection of 100 billion stars 80 000 light years in diameter. The galaxy is made more interesting by a slight bulge towards the centre

**Figure 3.1  Cosmic distances**

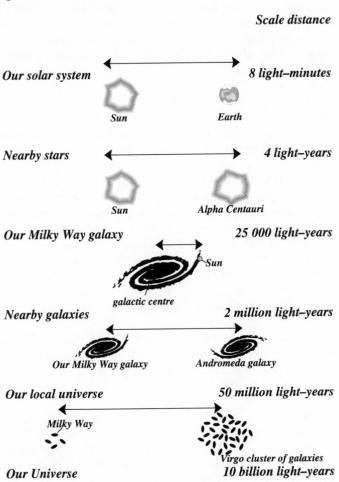

*Scale distance*

*Our solar system*                                    *8 light–minutes*

Sun          Earth

*Nearby stars*                                        *4 light–years*

Sun          Alpha Centauri

*Our Milky Way galaxy*                                *25 000 light–years*

Sun
galactic centre

*Nearby galaxies*                                     *2 million light–years*

Our Milky Way galaxy          Andromeda galaxy

*Our local universe*                                  *50 million light–years*

Milky Way
Virgo cluster of galaxies

*Our Universe*                                        *10 billion light–years*

These are measured in distances travelled by light in certain times. For instance, one light-minute is about $2 \times 10^{10}$ metres and one light-year is about $10^{16}$ metres.

and the four or five arms that give the galaxy its spiral. Astronomers have determined that the galaxy is rotating around its centre. The Sun makes one orbit every 200 million years, moving at an incredible 230 kilometres per second. The Milky Way is a very average-looking galaxy and contains stars in all stages of life. Most of the new stars are being formed in the arms of the spiral where there is still enough raw gas and dust to act as seed material. On the other hand, the centre of the galaxy appears overpopulated with old, redder stars, indicating that stars started forming in the centre earlier than in the outer regions.

In 1918, astronomers determined the position of the Sun and Earth within the Milky Way. Just a few years later they began to see the even bigger picture. In the early 1920s, the discovery of the 'redshift' of galaxies by American astronomer Vesto Melvin Slipher launched the study of the size of the Universe. At the time there was general agreement that most stars belonged to the one galaxy, ours. The small, faintly glowing patches in the sky that we now know to be other galaxies were also thought to be part of the Milky Way. The patches were called nebulae, and were lumped in with the nebulae left behind by supernova explosions and those associated with stellar nurseries. For example, our neighbouring galaxy Andromeda was known as the Andromeda Nebula, since its real distance was unknown.

But what is redshift? In 1842, the Austrian physicist Christian Doppler first described the effect that now bears his name. The familiar increase in pitch of a train whistle as the train approaches is a result of sound waves being compressed ahead of the moving source. The compressed sound waves have a shorter wavelength and a higher frequency or pitch. A train moving away will produce the opposite effect—the stretched out sound waves will have a lower frequency. The magnitude of the frequency change is directly related to the speed of the train. This is the 'Doppler effect', which is also observable with moving sources of light. Slipher searched for this effect in the light from glowing nebulae.

He used a spectrograph (essentially a glass prism) to break up the light from a distant nebula into its component wavelengths. He noticed that certain features of the spectrum were not at the expected wavelengths, those wavelengths regularly measured in the laboratory. They were shifted either towards the red end of the spectrum—'redshifted'—the increase in wavelength indicating the source of light is moving away, or towards the blue end—'blueshifted'—an approaching source. Slipher concentrated on nebulae situated away from the band of stars that defines the densest part of the Milky Way. His first target was the prominent Andromeda Nebula in which he detected a blueshift, while most of the other nebulae revealed redshifts. In general, Slipher found that the fainter nebulae were moving away, or receding, the fastest. In contrast, the nebulae within the densest part of the Milky Way had much smaller spectral shifts, and were just as likely to be redshifted as blueshifted.

The next insight came with the work of Edwin Hubble and his assistant Milton Humason at the end of the 1920s, also in the United States. Humason used the 2.5 metre diameter Mount Wilson telescope to photograph over 100 faint nebula over a seven-year period. Originally a mule driver who carted material up Mount Wilson during the construction of the observatory, Humason was eventually promoted to a janitor position and finally to a job as a telescope spectroscopist. His careful and painstaking work involved picking out faint nebula from crowded star fields and then positioning the entrance slit of the spectroscope over the nebula. Photographic exposures ranging from several hours to several nights were necessary for these dim fuzzy patches of light. Today, with the excellent tracking mechanisms in modern telescopes, this task is a cinch. However, the dedicated Humason had to continually check the telescope's alignment and make adjustments of the slit position throughout the long, cold nights. Luckily, all his work was not done in vain.

As well as measuring the velocities of nebulae using their spectra, Hubble and Humason were also trying to measure distances to the nebulae using special stars called Cepheid Variables. These stars vary in brightness in a regular way, and observations by other astronomers of close-by Cepheid Variables had shown that the distance to the star could be calculated by knowing the average brightness of the star and the frequency of the brightness variations. On the night of 6 October 1923, Hubble discovered a Cepheid Variable star in the Andromeda Nebula. Much to his amazement, he calculated that the distance to the Nebula was 1 million light-years (later measurements revised this to a little more than 2 million light-years), much greater than the size of the Milky Way which has a size of about 80 000 light-years. It was only then that astronomers realised that many of the nebulae were not local gas clouds but distant galaxies, some just like the Milky Way, some much smaller, some bigger. The Andromeda Nebula suddenly became the Andromeda Galaxy.

Hubble and Humason now had distances and velocities for a large number of galaxies. They found that the closer galaxies sometimes had redshifts and sometimes, like Andromeda, they had blueshifts. This indicated that local galaxies had a complicated motion, greatly affected by the gravitational pull of neighbouring galaxies. However, the more distant galaxies were *all* rushing away! Using the classic Doppler equation, the redshift gives the so-called recessional velocity, the apparent speed at which the galaxy is moving away from us. When Hubble plotted the recessional velocity of galaxies against their distance, he found a straight-line relationship now known as Hubble's Law. The more distant the galaxy, the faster it moved away. To his surprise, this was true in every direction and the recessional velocities could be very large. A galaxy at a distance of 100 million light-years was moving away with a speed of 1500 kilometres per second (or 5 million kilometres per hour)! Hubble had discovered the expansion of the Universe.

## The Big Bang theory

Why do the galaxies race away from us in every direction? Do we sit in a special place in the Universe? How big is the Universe? These are some of the questions raised by Hubble's work in the astronomical, and wider, communities in the decades after his discovery. Eventually, after many unsuccessful starts, one theory—the Big Bang—has survived that explains the expansion. The Big Bang theory also explains why the Universe contains so much helium and why we are immersed in a Universe-wide sea of microwave radiation.

The theory, first proposed by George Gamow, Ralph Alpher and Robert Hermann in 1946, has been refined by many who followed them. It starts with the premise that the Universe is not infinitely old. Today we believe the Universe started in an intense fireball just 10 to 20 billion years ago. The explosive nature of that event triggered the expansion, but not in the way that a familiar sort of explosion would. This might be a little hard to picture but it helps to think of the classic analogy used to explain the expansion of the Universe—an expanding balloon. Think of a balloon with coins stuck to the surface, each coin representing a galaxy. Here we confine the entire Universe to the *surface* of the balloon. (The main problem with this analogy is that we're representing a three-dimensional Universe by a two-dimensional surface!) As we inflate the balloon, the space between 'galaxies' increases. Imagine standing on one of the coins. You would see every other coin move away from you, and the more distant coins would move away *faster*. This, of course, is exactly the sort of thing observed by Hubble with real galaxies. Clearly, this behaviour would be the same if you had selected any other coin to be home. There is no 'special' galaxy, and there is no 'centre' of the Universe. The Universe has expanded from nothing, just like the balloon, and continues to grow through the expansion of space.

Some scientists dared to think about what happened before the Big Bang. The X-ray astronomer Herbert Fried-

**Figure 3.2 Big Bang expansion**

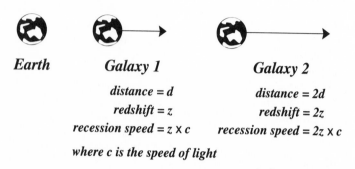

*Earth*      *Galaxy 1*          *Galaxy 2*

*distance = d*              *distance = 2d*
*redshift = z*              *redshift = 2z*
*recession speed = z × c*   *recession speed = 2z × c*

*where c is the speed of light*

As we view galaxies at greater distances their 'redshifts' (or speeds of recession from us) increase in proportion to their distance. Doubling the distance gives double the redshift. We can say that 'recession speed = H × distance'. H is a constant number known as Hubble's constant. The exact value of H is not agreed upon but it is something like 20 km/sec/million light-years.

man once reminded us of a comment by St Augustine who had asked, 'What was God doing before he created Heaven and Earth?'. The early leader of the Christian church then answered his own question with, 'He was preparing a Hell for those who inquire into such high matters.'! The scientific answer is that it doesn't make sense to ask what happened before the Big Bang. Time did not exist then, just as space as we know it did not exist. The clock of history started at the same instant that space started to expand from the infinitely small and dense fireball. We can measure the current rate of the expansion of the Universe and calculate how long ago that instant was.

After all, if we know how fast our galaxy and another galaxy are rushing apart, and if we know their separation, then we can estimate how long ago they existed at the same point. While it's straightforward to calculate the recessional velocity of that second galaxy from a spectral measurement, measuring its distance is more difficult. This is especially true for very distant galaxies, where it's impossible to pick out useful stars like Cepheid Variables. This obviously causes an uncertainty in the plotting of the classic Hubble diagram of recessional velocity versus distance, and in the current expansion rate. However, keeping these difficulties in mind, it's possible to make some good estimates of this number. Astronomers have calculated an age for the Universe of somewhere between 10 and 20 billion years. To put that in some sort of context, we believe that the Sun and our Earth are about 4.5 billion years old.

The finite age of the Universe solves a very old problem in astronomy known as Olber's Paradox. Put simply, the paradox revolves about the question 'Why is the sky dark at night?'. As we will see, the night sky should not be dark if the Universe is infinitely large and infinitely old. Named after the nineteenth-century German astronomer who revived discussion about it in the 1860s, Olber's Paradox was probably first raised back in the time of Newton and Kepler in the seventeenth century.

Isaac Newton was a great believer in a Universe that was static and infinite in extent—a popular view of the time. Such a view was necessary, he thought, given his new theory of gravitation. If the Universe was not infinitely large, there would be a centre and an edge to all of the matter, and gravity would tend to attract the matter towards the centre. A single consolidated mass would be the result. This rather uncomfortable situation would be solved if every star experienced an equal gravitational pull in all directions.

The famous seventeenth-century astronomer Johannes Kepler, who actually died a few years before Newton's birth, had a different view. If the Universe was infinite in extent, Kepler reasoned that we would see a star in every part of the sky. It would be impossible for us to find a dark gap and the night sky would be very bright indeed. That the sky is dark at night is the paradox. Kepler used the paradox to argue that the Universe was not infinitely large. However, others, including Olber in the 1860s, were of the opinion that light from very distant stars was obscured by regions of dust in space. The Big Bang concept provides a different and very simple explanation. If the Universe is only 15 billion years old, then we aren't able to see stars more distant than 15 billion light-years. Given the finite speed of light, the light from these stars just hasn't had time to reach us. Think of the balloon analogy again—there may be galaxies on the surface of the balloon more distant than 15 billion light-years, but we can't see them. The sky is therefore dark at night. The Universe will need to be a lot older before we can start reading newspapers outside on moonless nights. Calculations have shown that we would need to see stars out to a distance of 1 trillion trillion light-years for the night sky to become day!

## A cooling remnant of the fireball

At the time of its proposal in 1946 and for the next twenty years, the Big Bang was simply one of many cosmology theories. However, in the mid-1960s it sprang to prominence with the discovery of remnants of the primordial fireball in the form of a weak and all-pervading radiation field. Surprisingly, this weak radiation has an enormous influence on the most powerful cosmic rays as we'll see later in the book.

In 1965, Arno Penzias and Robert Wilson of the AT&T Bell laboratories in the US were using a huge microwave antenna to study sources of radio interference from the sky. Their plan was to use this antenna system to relay television and radio signals to a satellite for transmission across the Atlantic ocean. As part of the work, Penzias and Wilson upgraded the antenna for a very sensitive study of radio signals from the Milky Way, as this background could be a threat to the communications mission. They found a persistent hiss in their receiver present in every direction, a signal from microwaves with a wavelength of 7.35 centimetres. They tried everything to get rid of this noise. Originally, a problem with the radio receiver was suspected—what other explanation could there be for a signal that appeared to be coming from everywhere? After cooling parts of the receiver with liquid helium and cleaning pigeon droppings from the inside of the antenna horn (stuff referred to by Penzias as 'suspicious white dielectric material'), the annoying hiss was still present.

The only conclusion possible was that this radiation was real, and it was filling the sky. The two Bell Lab scientists characterised the radiation as having a characteristic temperature of somewhere between 2.5 and 4.5 degrees on the Kelvin scale. In other words, this is the sort of radiation you would see emitted from an object with a temperature just a few degrees above absolute zero ($0^\circ$ K = $-273$ $^\circ$C, the minimum possible temperature). Though very cool, this radiation is copious. Penzias and Wilson calculated that a person standing outside will have 1000 trillion of these microwave photons striking their head every second!

In that same year, 1965, a group of theoretical physicists was working just up the road from the AT&T antenna station, at Princeton University. Led by Robert Dicke and James Peebles, the group was interested in the Big Bang model and had been considering experimental tests that could verify or shoot down the theory. Gamow and other originators of the theory had realised that remnants of the hot fireball at the beginning of the Universe might be detectable today. When the Universe was very small and very dense, the temperature was so high that the photons flying around were in the X-ray and gamma-ray range of the electromagnetic spectrum. Since that time, space has stretched as the Universe expanded, just like the surface of the balloon in our analogy. One result of this stretching has been that the wavelengths of the light photons have been pulled longer. Longer wavelengths mean the photons have smaller energies and a lower temperature.

Dicke and Peebles heard about the discovery of Penzias and Wilson and made the connection straight away. Their calculations had predicted that the remnant radiation from the Big Bang would now have a characteristic temperature of just a few degrees above absolute zero. Almost immediately, two papers, one by Dicke and Peebles and one by Penzias and Wilson, were published in the same issue of *The Astrophysical Journal* in 1965. News of the discovery also made it to the front page of the *New York Times*—Penzias and Wilson said that it was only then that they realised the significance of their observation! They were rewarded more tangibly in 1978 with the Nobel Prize for physics.

This discovery of the microwave background radiation promoted the Big Bang model from one of two or three competing cosmology models to the premier one. The existence of the wispy handprint of the original fireball gave enormous weight to the theory, and it's the fireball that solves the problem of too much helium in space, a problem first realised by astronomers in the 1940s and 1950s. Stars certainly make helium out of hydrogen in their cores through the process of nuclear fusion, but there is too much helium around to be explained by this mechanism. At the beginning of the Universe, the temperature and density were similar to the conditions we know exist inside stars today. If helium is produced in stars, it must have been produced everywhere in the hot early Universe, together with traces of the next heavier elements, lithium and beryllium. These elements have since expanded with the rest of the Universe, seeding the first generation of stars and galaxies.

So the Big Bang Theory has successfully explained, or predicted, the three main pillars of modern cosmology—the expansion of the Universe, the proportions of the light elements, and the microwave background radiation. Over the years, the model has gone from strength to strength with more confirming observations, including the precise measurement of the microwave background by the COBE (COsmic Background Explorer) satellite in 1992. These observations showed that the radiation followed the exact spectrum predicted by the theory and was incredibly uniform over the sky. The radiation is now known to have a characteristic temperature of $2.7^{\circ}$ K, right in the ballpark of the original estimate of Penzias and Wilson.

## Quasars and the active galaxies

The Universe is a big place. It's hard enough to come to grips with the scale of the Milky Way galaxy, our collection of 100 billion stars that takes light 80 000 years to cross from one end to the other. However, astronomers estimate that there are tens of billions of *galaxies* in the observable Universe! They range in size from small irregular clumps of stars like the Magellanic Clouds, through to spiral galaxies like our own, to giant elliptical galaxies ten times larger than the Milky Way. Apart from size, we can classify galaxies in terms of their powerful emissions—some galaxies seem to be much more than collections of billions of ordinary stars. Their enormous energy outputs lead many to suspect that they harbour exotic super-massive black holes at their centres, the huge gravitational force of the black hole being the most obvious source of the energy. We know that these galaxies output huge amounts of radiation, ranging from radio waves to gamma-rays. More importantly for our story, we suspect that these objects are responsible for the production of the highest energy cosmic rays. We'll describe more about these black hole 'engines' later, but first we'll follow the story chronologically. The discovery of the most amazing of all galaxies, the quasars, was another highlight of the decade of the 1960s.

In 1960, Alan Sandage presented his observations of a radio star called 3C–48 to a meeting of the American Astronomical Society. This 'star' was one of the strong radio sources listed by Cambridge University astronomers in their third catalogue of radio sources (3C stands for the third Cambridge catalogue). Sandage was the first to identify a source of visible light at the site of the radio signal (its so-called optical counterpart) and was puzzled by what he saw. Unlike many of the 3C objects, this did not look like a galaxy. Instead, the photographic plates showed a star-like object with a very puzzling spectrum including very strong emission lines—lines that Sandage could not recognise as coming from any known element or compound! He placed this puzzle in the too-hard basket and it was to stay there for two years.

In the meantime a group of Australian radio astronomers led by Cyril Hazard made observations of another source from the catalogue, 3C–273. No optical counterpart had been found for this source, since the sky coordinates of the radio signal couldn't be accurately determined by the early radio telescopes. It so happened that in 1962 the Moon passed in front of 3C–273, and Hazard and his colleagues were ready, tracking the strength of the radio signal as it did so. The precise time of this 'occultation' gave a very accurate position for the source and also showed that the source actually had two 'cores' or points of emission. One core was stronger than the other with a separation of a very tiny $\frac{1}{200}$ degrees. Hazard and his team carefully examined the photographic survey plates of that region of the sky. They found a very faint blue star in the position of the weaker of the two radio cores.

Hazard then asked the American astronomer Maarten Schmidt to train the 200-inch Palomar optical telescope onto this star. Schmidt was able to identify a faint jet of light at the position of the other, stronger radio core. This was turning out to be a very strange star! It got even stranger when Schmidt took a spectrum of the light. He found, like Sandage had two years before, a set of emission lines that he couldn't identify. There was something vaguely familiar about the spectrum though and the penny finally dropped in Schmidt's mind six weeks after his observation. He realised that the pattern of spectral lines was just like the pattern in the laboratory spectrum of hydrogen, except that the lines were in the wrong places! The wavelengths were all shifted towards the red end of the spectrum by 16 per cent. Schmidt realised the implication of his discovery immediately and is said to have gone home and announced to his family that 'something really incredible happened to me today'.

Schmidt recognised that the redshifted spectrum was caused by 3C–273 rushing away from the Earth at an enormous speed. The 'star' was not a star in our galaxy but most probably a very distant galaxy, with an enormous recessional velocity caused by the expansion of the Universe. The redshift of 16 per cent implied a velocity of 16 per cent of the speed of light, or 48 000 kilometres per second! The magnitude of the redshift was much bigger than had been seen before—for example, Hubble and Humason had only seen redshifts up to around 1 per cent. Schmidt's colleague, Jesse Greenstein, soon realised that the earlier mysterious spectrum of 3C–48 measured by Sandage was in the same class but even more extreme—the shift in the spectral lines was a colossal 37 per cent. It was little wonder that nobody had recognised the pattern earlier. Yet the implication of that recessional velocity was startling. Given Hubble's expansion law (relating recession speed to the object's distance), this velocity implied that the source was 4 billion light-years away. How could a galaxy be so distant and also appear as one of the brightest radio sources in the sky?

3C–48 was seen to vary in radio brightness over periods less than a day. This simple observation had an unbelievable ramification since there's a rule in astrophysics that says that an object can't vary in brightness faster than the time it takes light to cross the object. It might help to think about it this way. Imagine some sort of astronomical object, say one with a diameter of 10 light-days, and suppose that radio waves are *simultaneously released* from all points on the object. Imagine that the object is transparent so that we can see radiation released from the object's far side. Because of its size, radio waves from the far side of the object will arrive at Earth ten days later than the waves emitted from the side of the object closest to us. In other words, even if the object released a very short pulse of radiation (say 1 second wide) from every part of itself, we would see the pulse lasting ten days! If the object released a pulse longer than ten days, we would see its true duration, but no pulse duration shorter than ten days could be seen because of the object's 10 light-day size. Consequently, the variations in brightness of 3C–48 over a time scale of a day indicated that the object's emission region was incredibly small, the order of a light-day. This is a region not much larger than our solar system!

Greenstein and Schmidt coined the term 'quasars' or quasi-stellar objects for these phenomenal sources. Since the early 1960s astronomers have discovered hundreds of such objects, all characterised by large radio brightnesses and huge distances. Some of these have been observed at distances exceeding 10 billion light-years and it's estimated that each has a brightness equivalent to hundreds of galaxies. How an object can shine with that sort of energy from a volume the size of our solar system is the challenge that has faced astrophysicists for more than thirty years.

The evidence seems to point to quasars being *strange* galaxies with powerful 'engines' at their centres. These engines are clearly sources of many types of emission. Several quasars have been observed right across the electromagnetic spectrum, from radio waves through to gamma-rays. Importantly, from our point of view, current models for quasars also point to these objects as powerful particle accelerators. Quasars are certainly very different from galaxies like the Milky Way. Our galaxy certainly produces a lot of radiation, but at nowhere near the level of the quasars. There is a class of galaxies that fills the gap in power output between the quasars on the one hand and galaxies like the Milky Way on the other. These are the so-called 'active' galaxies and while we can't see them at the same distances we see quasars, they are remarkable enough in their own right.

The American astronomer Carl Seyfert discovered the first active galaxies during World War II when he was compiling a catalogue of spiral galaxies. He stumbled upon a subset of these galaxies, now known as Seyfert galaxies, which showed very bright and compact central cores. Seyfert's spectral studies revealed that much of the brightness of one of these galaxies comes from large turbulent clouds of extremely hot gas moving rapidly in its central regions. A few dozen galaxies are now known to be Seyfert galaxies and the most luminous in the class approach the energy output of weak quasars.

Another class of active galaxies is represented by the famous galaxy Centaurus A, one of the brightest radio sources in the sky and one discovered by early Australian radio-astronomers. It was almost the first radio source discovered when radio telescopes began operation worldwide in the 1940s. Part of the reason for the brightness of this galaxy is its closeness—at a distance of only 15 million light-years it lies in our 'backyard'. Cen A is striking enough through an optical telescope with the glow of the giant elliptical galaxy partly obscured by a thick belt or 'lane' of dust across its centre. However, the true picture is only seen in the radio spectrum. Here we see two extremely strong sources of radio waves on either side of the dust lane, called radio lobes, with a second pair of lobes further out. The outer lobes are actually a distance of 1 million light-years from the centre of the galaxy and Centaurus A takes up a whopping 4 degrees in the radio sky! Cen A was the first known member of our second active galaxy class, the so-called double radio sources. Radio astronomers have discovered much structure in the lobes of many of these galaxies with high intensity 'knots' and lower intensity voids. Like some of the closer quasars, some of these double radio galaxies are observed in other wavelength bands including X-rays and gamma-rays.

## Active galaxy engines

No doubt something special exists in the central regions of quasars, Seyfert galaxies and double radio sources. The central regions have recently been given a name of their own, Active Galactic Nuclei or AGN. What is the mechanism in operation inside these AGN? If we define 'efficiency' in terms of how much energy we can get out of a process for a given amount of fuel going in, then one of the most efficient processes is nuclear fusion. This is the energy source of normal stars and a topic we'll be discussing in the next chapter. Fusion gets an efficiency rating of 0.7 per cent when the energy equivalent of the fuel's mass is included. That fraction of the mass of the stellar nuclear fuel is ultimately converted into energy. Given all the mass available, this small efficiency actually produces a huge amount of energy and this is one reason that scientists are spending billions of dollars trying to replicate stellar fusion for power stations. However, there are processes in astrophysics that are at least twenty times more efficient, though quite impractical for power generation on Earth. Let's start with an example involving a neutron star.

A neutron star is a very compressed object that results from the death of a massive star in a supernova explosion. The incredibly compact core of the star is left behind, typically with a mass a little larger than our Sun, squashed in a volume only 30 kilometres across. This means that the density of the neutron star is extraordinary, akin to the density of the nucleus inside an atom. Now, consider a neutron star and a normal star in orbit about each other. If the orbit is small enough, gas from the outer atmosphere of the normal star will be attracted by the strong gravity of the neutron star. The strong gravity results from the large mass of the neutron star concentrated in so small a volume. The captured gas will collect in an 'accretion' disk. This disk lies in a plane at right angles to the rotation axis of the neutron star. As the mass spirals towards the very compact object it will gain energy, just as a falling ball gains speed as it falls towards Earth. The source of the energy in both cases is gravitational and since the gravity is so strong around a neutron star the energy gained by a falling gas atom is enormous. The energy often manifests itself as heat and it's the extremely hot inner edge of accretion disks that gives rise to X-ray emissions in these binary star systems. The efficiency of the process is impressive—it releases an amount of energy equivalent to around 20 per cent of the mass of the falling gas.

This is the sort of efficiency needed in an active galactic nucleus. While there have been suspicions about the nature of AGN engines for many years, recent observations by the Hubble Space Telescope have provided some confirmation. It seems that AGN harbour super-massive black holes and their energy comes from the gravitational pull of

**Figure 3.3 Accretion around a neutron star**

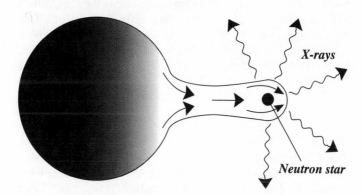

If a neutron star is formed in a binary system, it may be so close to the other, normal, star that its gravitational pull may attract material from the outer regions of that star. This material will heat up as it approaches the neutron star and form a hot 'accretion disk' in which there are strong electric and magnetic fields capable of accelerating cosmic ray particles. An example of such a system, Cygnus X–3, will be discussed in Chapters 7 and 8.

those objects. In other words, the central regions of AGN may well be scaled-up versions of our accreting neutron star—instead of a neutron star, a giant black hole might gather a huge accretion disk which will produce radiation through this very efficient mechanism. Of course, a black hole is an object with an extreme gravitational field, so strong that not even light can escape from its environs. Astronomers predict that individual black holes result from the collapse of very massive stars. However, the black holes thought to exist in the centres of AGN are much larger—instead of a mass of five or ten solar masses, they have the mass of a million or more Suns.

M87 is a giant elliptical galaxy near the centre of the Virgo galaxy cluster 50 million light-years away. As the brightest radio galaxy in this constellation, it's also known as Virgo A. Astronomers have known for many years that this object is a double radio source. In visible light it exhibits a jet shooting out 5000 light-years from its nucleus. This faint blue feature is a tiny version of the jet emanating from 3C–273 observed by Maarten Schmidt—in that case the jet is estimated to be 160 000 light-years long. Early optical observations of M87 also showed another feature common to quasars—its stars are quite strongly clustered towards the centre of the galaxy, giving the object a very bright core. A closer examination of this structure had to wait until 1994, when the repaired Hubble Space Telescope peered into the centre of M87. Holland Ford and Richard Harms, two astronomers working on the observations, were astonished at the clarity of the pictures they saw. They discovered a disk-like whirlpool of hot gas orbiting the nucleus. Given the elliptical nature of the galaxy as a whole, it was somewhat of a surprise to discover this spiral structure near its centre. The excellent resolution of the Hubble telescope allowed Ford and Harms to make spectroscopic measurements of the inside edge of the whirlpool. Their aim was to use the Doppler effect to reveal the speed of the swirling gas and dust. The gas appeared extremely hot, about 10 000 K, and the light emitted was redshifted on one side of the disk, and blueshifted on the other. This was exactly what the astronomers would expect for a rotating disk viewed at an angle with one side of the disk moving away and the other side approaching.

It was the astounding magnitude of the speed that got Ford and Harms excited—a huge 2 million kilometres per hour, or 55 kilometres every second! Here was the evidence for a black hole. The speed of the spiralling gas provides rather a direct measurement of the mass contained at the centre of the whirlpool, using a law known since the time of Kepler in the seventeenth century. The conclusion of the observation was that a mass equivalent to 2.4 billion Suns is concentrated in a space not much larger than our solar system. This didn't leave much doubt about the nature of the central object! In the words of Holland Ford, 'if it isn't a black hole, I don't know what it is. A massive black hole is actually the conservative explanation for what we see in M87. If it's not a black hole, it must be something even harder to understand with our present theories of astrophysics.' This observation was not a fluke—in December 1995 the same team found another super-massive black hole in the core of the active galaxy called NGC 4261. This galaxy is also located in the Virgo constellation, but twice as far away at a distance of 100 million light-years.

It might be tempting to consider super-massive black holes more unlikely, or more exotic, than black holes predicted as the endpoint of very big stars. It turns out that the conditions required for the creation of the super-massive variety are far less demanding. The General Theory of Relativity predicts that the density of matter required to create a black hole (in order for gravity to be strong enough to stop light escaping) is inversely proportional to the square of the mass of the hole. So while it takes a density of 10 billion billion kilograms per cubic metre (twenty times the density of an atomic nucleus) to produce a black hole with the mass of the Sun, the density required to produce a billion-solar mass hole is only 10 kilograms per cubic metre—one hundred times *less* dense than water!

So the favoured model of an active galactic nucleus starts with an enormously massive black hole surrounded by a whirling accretion disk that funnels matter inwards. That matter might be gas and dust from the interstellar environment as well as whole stars! The matter in the accretion disk moves faster and faster as it spirals inwards. The material also gets hotter as friction converts some of the energy of motion into heat. The forces acting on the gas at the inner edge of the disk will be enormous. Apart from the pressure due to the outflow of heat and the pressure from the very intense radiation, there is a very strong centrifugal force acting on the rapidly spinning material. Because of these factors, the black hole won't swallow all of the matter. In fact, for some of the material, the path of least resistance is not inwards into the hole nor outwards in the plane of the accretion disk. Instead, this material shoots out in two oppositely directed jets perpendicular to the disk at speeds approaching the speed of light.

These relativistic jets give rise to much of the radio emission we see from quasars and active galaxies. Apart from high-speed particles, astronomers believe that jets also squirt out fast-moving magnetic fields. The combination of these fields and particles results in radio emission via the so-called synchrotron process. Energetic charged particles (principally electrons) lose energy as they move in spiral paths around the magnetic field lines and that energy appears mainly as radio waves. Because of the high speeds involved, the jets remain very tightly bundled and extend for many thousands of light-years from the galactic core. The structure of the tangled magnetic field in the jets is constantly changing as the core ejects more and more material. On the other hand, it seems that semi-permanent knots or concentrations of the field may form, even at large distances along the jet. Also, shock waves constantly move up the jets from the violent core. As shock waves meet knots in the magnetic field, astronomers believe that very energetic particles carried by the shock front interact with the intense magnetic knot producing very strong synchrotron radiation. We see the radiation from these jets as the 'lobes' in double radio sources.

Recently, a Dutch astrophysicist named Peter Barthel has put forward a popular theory on the unification of the various classes of active galaxies. Barthel recognised that quasars and active galaxies all require the sort of power provided by a super-massive black hole and an accretion disk. He has proposed that quasars and the other classes of active galaxies are really all the same type of object and that the classification we give to each object depends on the angle of our view! As in the standard theory, the central black hole accretes a massive spinning disk and shoots out a pair of energetic jets. However, if the orientation of the object in the sky is such that we view the accretion disk nearly edge on, we would get a very good view of the knots in both jets. Astronomers would classify this object as a double radio source. On the other hand, Barthel argues, if one of the jets happens to be pointing close to our direction, we get a very different picture. Then the enormous amount of radiation from the jet would give the characteristic bright, star-like appearance of a quasar. Barthel's model is very appealing. It fulfils one of the missions of scientific endeavour in that it attempts to simplify the enormous complexity of Nature by searching for a unifying concept. While Barthel's model is in its infancy, no-one has yet raised serious objections.

Active galaxies and the massive black holes that probably power them are prime sites for accelerating the highest energy cosmic rays. As we'll see later in the book, these are almost the only astronomical objects capable of giving a proton over 50 joules of energy—one hundred million times more energy than we can give to protons in earthbound particle accelerators. Once produced, the highest energy cosmic rays barrel through intergalactic space, their only barrier being the photons of the microwave background radiation. Collisions with these wispy remnants of the Big Bang rob the cosmic rays of some of their energy and provide us with some valuable clues to their origin.

# CHAPTER 4

# STARS AND THEIR FATE

We're now going to zoom in towards our more local environment. Our own Milky Way galaxy contains many objects important to our survey of high energy astrophysics. Many of the high energy phenomena we'll meet in this book result from the final death throes of stars somewhat more massive than our Sun. Astronomers believe all stars begin life in the same way as tenuous clouds of gas and dust that collapse under gravity. However, their initial mass is the property with the major influence on how the stars evolve, how long they live, and the nature of their death. A massive star will end its life in a cataclysmic supernova explosion, leaving behind a remarkable neutron star (or perhaps a black hole) which in turn will produce many of the phenomena seen on the cutting edge of high energy astrophysics.

## The birth of stars

We have a general picture of star-birth as a collapse of a slowly spinning ball of gas and dust. The details of this process aren't well known, and in particular there's no clear theory about planet formation in the latter stages of the collapse. However, studies of these processes are becoming easier with the advent of large infra-red telescopes capable of seeing into the murky parts of our galaxy that contain stellar nurseries, regions dense with gas and dust. The density of the predominant gas, hydrogen, is typically about

100 atoms per cubic centimetre, a hundred times larger than the average density in the galaxy. It exists in large clouds, together with dust and other gas formed in earlier generations of stars. Perhaps seeded by a particularly dense region, the cloud collapses under the force of gravity to produce a spinning ball of tenuous gas called a protostar.

The protostar won't immediately look like or behave like a sun. It isn't yet producing energy via nuclear fusion, but it does start to get hot. The gas heats up through a process known as Kelvin–Helmholtz contraction. In the same way a bicycle tyre heats up as we increase the pressure of the air inside it, the temperature of a protostar continuously increases as the gas collapses.

Two British astronomers, Arthur Eddington and Robert Atkinson, realised in the 1920s that if the density and temperature climbed to high enough levels in the central core of the protostar, it would be possible for the nuclei of two hydrogen atoms to collide with such a force that a nuclear reaction would occur. The temperature required is in excess of 10 million degrees Celsius. At this temperature the atoms of hydrogen (each consisting of a proton nucleus with an orbiting electron) are stripped of their electrons. The protons, having high speed as a consequence of the enormous temperature, now collide with such a force that they overcome the extremely strong electrical repulsion between two positively charged particles. The result of this collision is a fusion reaction which proceeds along a three-step process involving other collisions to produce a nucleus of helium. The process releases energy, explained by Einstein's relativity theory, because there's a minute difference between the mass of the nucleus of a helium atom and the mass of the four protons that formed it.

## Stellar middle age

The onset of nuclear fusion marks the transition from a protostar to a star. Our Sun made this transition approximately 4.5 billion years ago, and is still fuelled by its massive reserves of hydrogen. Its power output of over $10^{26}$ watts

55

means that the Sun is losing hydrogen at a rate of about a million tonnes per second. This sounds a lot until you realise that the mass of the Sun is currently 2000 trillion trillion tonnes! Astronomers believe that the Sun will consume its hydrogen fuel for another 5 billion years before moving on to the next stage in its evolution. We can only hope that by that time our descendants have perfected the science of interstellar travel! The approach of the end of the hydrogen fuel at the Sun's core will mark the end of any life in our solar system.

The stellar fusion processes forge many of the atomic elements we're familiar with in everyday life. Indeed, these furnaces are the only sources of these elements. The Big Bang theory tells us that only the lightest elements, hydrogen and helium, were created at the beginning of time. The heavier elements, those that go into making our Earth, our atmosphere and our bodies were formed later inside stars and during the cataclysmic stellar explosions called supernovae. These explosions distribute this newly formed material around the galaxy, continually enriching the interstellar medium with heavy elements. It may be trite to say but there's no doubt that we are made of stardust.

The more massive a star, the more quickly it will consume the hydrogen fuel at its core. This is because the central pressure (and hence the temperature) is much larger in these stars, allowing a very rapid fusion reaction. These stars are the first to move onto the second stage of stellar evolution, becoming red giants, when the last of the hydrogen at the stellar core is consumed. At this stage, the stellar core consists of helium, the product of hydrogen fusion. Helium itself can be the fuel of a different fusion reaction where carbon and oxygen are formed and vast amounts of energy are again released. However, these reactions require even greater core temperatures and these conditions don't exist until the end of hydrogen fusion. The transition from a star fuelled by hydrogen to one fuelled by helium happens in a very short time after the hydrogen runs out. It's then that the star changes its appearance in a striking way.

## Stellar old age—red giants

Just before this transition a delicate balancing act is in operation, a balance that's existed since the star first produced energy through fusion. The huge inward gravitational force has been balanced by the thermal pressure exerted outwards by the furnace at the stellar core. However, when the star exhausts its supply of hydrogen fuel, the balance is ruined and the gravitational force is free to collapse the star. Luckily, the collapse is not catastrophic because the collapsing star causes the core to become hotter (the Kelvin–Helmholtz mechanism again) resulting in temperatures high enough for helium fusion to commence.

The rekindling of the stellar furnace, now with a different fuel, actually reverses the stellar collapse. Helium fusion produces more energy output than the previous hydrogen reactions and the balancing act between gravity and the thermal output results in a new stable size for the star. The star is now truly giant. It is producing more energy overall but now has a huge surface over which it can radiate this heat away. This results in the slightly surprising fact that, despite a much more vigorous core, the surface temperature is now cooler than before. The star now appears red in colour. This red giant star is again stable and may happily fuse helium for a billion years.

The atomic products of helium fusion include carbon, nitrogen and oxygen. These elements build up in the core of the red giant star and may become, in time, the new stellar fuel after the helium reserves have been exhausted. In fact, the most massive stars will go through a series of stages where successively heavier nuclear fuel fuses together to produce heavier and heavier elements. Each stage requires the core temperature of the star to be even higher than before in order to overcome the reluctance of the heavier elements to fuse together. In turn, these higher temperatures speed up the burning process so that the time spent in each successive stage becomes shorter and shorter.

As an example, let's consider the life cycle of a star twenty-five times more massive than the Sun. Because of

57

the enormous pressure inside this huge object, even the hydrogen burning stage passes within a very quick 7 million years. Then 700 000 years of helium burning is followed by 600 years of carbon burning, one year of neon burning, six months of oxygen burning and one day of silicon burning! The final elemental product of this chain of events is the large and stable nucleus of iron. The fusion process stops here because iron is the most stable nucleus we know and isn't able to act as fuel for either a fusion or fission reaction. The iron core is incredibly hot and dense. Confined to a volume about 1000 kilometres in diameter at a temperature of a few billion degrees, the iron is a billion times denser than water!

## Star deaths and supernovae

The controlled sequence of fusion processes, from hydrogen to silicon fusion, releases enormous amounts of energy throughout the life of the star. At the same time it produces many of the elements necessary for life on Earth. However, the power of these processes pales into insignificance against the ferocity of the final death throes of a massive star—the supernova explosion. With an inert iron core, the bloated red giant star has no source of energy at its centre and for a final time the gravitational collapse of the star begins. In this case there is no new fusion reaction to stop the collapse and restore balance to the star. Instead the collapse continues, leading to a very rapid succession of events within seconds of the furnace switching off. This supernova releases energy in many forms including those accelerated charged particles we know as cosmic rays.

The iron core collapses under enormous pressure. Protons and electrons are squashed together to form neutrons, releasing as a by-product trillions of neutrinos. Typically, the core compacts down to a volume 30 kilometres in diameter with a density of about a million billion kilograms per cubic metre, the same density seen inside the nuclei of atoms! The outer layers of the star follow the collapse, moving inward at high speed. This material collides with the solid neutron core, becomes very hot and 'bounces' outward again. The bounce is driven in an explosive way by a combination of the extreme temperature of the gas and the pressure of the neutrinos escaping from the core. The shock wave moves out at very high speed, sweeping up more gas in the process and increasing the temperature of the material to extreme levels. Temperatures get so high that fusion is ignited in the hydrogen and lighter gases of the star's outer atmosphere. With an enormous explosive flash that lasts about a second, huge amounts of gas are fused and the supernova becomes as bright as a hundred billion stars!

After the cataclysmic explosion, the star continues to shine as brightly as 200 million suns for another two or three weeks as the supersonic shock wave continues outwards into interstellar space, heating up more and more gas. Being so bright, it is sometimes possible to view a supernova explosion with the naked eye. In 1054, the Chinese astronomer Yang recorded observations of what he called a 'guest star', a star that appeared out of nowhere and was bright enough to be seen during the day for a whole month. We now know that this was the supernova explosion that created the Crab Nebula in the constellation of Taurus, one of the most studied objects in modern astronomy.

A more recent example occurred in the early hours of 24 November 1987. A young Canadian astronomer, Ian Shelton, was taking a break from observing at a large telescope in southern Chile and was walking about outside the telescope dome. He noticed a 'new' star in the Large Magellanic Cloud, our small neighbouring galaxy. This object, named SN 1987A, was another of the rare, naked-eye supernovas, and the brightest one to appear in almost four hundred years. We know that this explosion produced neutrinos because two underground detectors on Earth detected a handful of these elusive particles. (Neutrinos are hard to detect because they don't readily 'interact' with matter. Most of the supernova neutrinos that reached Earth simply passed through the entire planet without any trace.)

Almost ten years after this explosion, we still don't know what has been left behind. The shell of exploding material is still hurtling away from the explosion site, blocking our view of what lies inside. We suspect that the supernova shock waves have accelerated particles to high energies, producing cosmic rays and other radiation. We won't know for sure until the exploding shell becomes thin enough for this radiation to escape. On the other hand we know very much more about the remnant of the Crab event seen by the Chinese nearly a thousand years ago.

## The embers of dead stars

Today, the area of the Crab explosion still glows. At the centre of the glowing Crab Nebula lies the rapidly spinning core of the long-dead star, an object known as a pulsar. This object produces a range of strong emissions, from radio to very high energy gamma-rays. What exactly is a pulsar and how does it produce this incredible range of radiation? Their discovery, and our understanding of their nature, provide an interesting case study in the practice of science and one that led to a Nobel Prize for one of the discoverers in 1974.

The story actually began many years earlier in the 1930s, soon after the theory of stellar fusion reactions had been proposed. Astronomers had observed supernova explosions and they were recognised as the endpoints of the lives of massive stars. But what became of the core of the star? The popular theory was that the core became a white dwarf star. Such stars were known to exist and were characterised as dim stars with white-hot surfaces.

The theory behind white dwarf stars was proposed by a brilliant young Indian astrophysicist, Subrahmanyan Chandrasekhar, in 1931. The twenty-one-year-old was on a boat journey from India to Cambridge University to take up studies for a doctorate in physics. During the long voyage, Chandrasekhar determined the characteristics of what he thought to be the ultimate form of a collapsed star. He proposed that so-called 'electron degeneracy pressure' halts the stellar collapse. In other words, the core, composed of nuclei of heavy elements immersed in a sea of swarming electrons, becomes stable when the gravitational collapse squeezes the electron 'gas' into its smallest possible volume. The pressure exerted by the compressed electron gas will support the weight of the core and a stable white dwarf star is formed. Chandrasekhar's calculations showed that a white dwarf would be extremely dense, with a mass of the Sun squeezed into an object the size of our Earth. Despite the weirdness of the object, and despite opposition from Arthur Eddington, the greatest astrophysicist of the time, the concept of white dwarf stars was eventually accepted.

Theories that proposed stars with densities higher than white dwarfs were considered unnecessary for thirty or so years after Chandrasekhar's theory was published. Most astrophysicists were happy with such stars being the remnants of supernova explosions. However, Chandrasekhar's theory did contain a prophetic conclusion. He found that the electron pressure would not be capable of stopping the collapse of a stellar core if the mass of that core was larger than 1.4 solar masses. In other words, a more massive core would collapse so rapidly under the huge gravitational forces that even the electron gas pressure would be overcome. The fate of such a collapsing core wasn't known and most astronomers moved on to other problems.

## Neutron stars and pulsars

Two exceptions were Fritz Zwicky and Walter Baade who had witnessed the publication of Chandrasekhar's work in 1932 and in the same year had seen James Chadwick's paper on his discovery of the neutron. Within months, the two American astronomers had come up with the theory of a 'neutron star'—a star composed entirely of neutrons with a density similar to the density of an atomic nucleus. They figured that if the mass of a stellar core was too large to be supported by electron degeneracy pressure, it would continue the collapse until it was stopped by a new outward force—that provided by *neutron* degeneracy pressure. The final state of the core would be determined by how far a star made of neutrons could be squeezed. Zwicky and Baade calculated that a neutron star with a mass 1.5 times larger than the Sun's would be tiny, with a diameter of only 30 kilometres. Such a crazy object was not attractive to the wider astronomical community though some theorists continued work on neutron stars during the 1930s including Robert Oppenheimer, who of course went on to become the chief scientist of the Manhattan Project that developed the atomic bomb. However, the theory lay dormant for thirty years until observations forced astronomers to dust off Zwicky and Baade's visionary work.

In 1967, a 24-year-old student named Jocelyn Bell was studying for a doctorate in astrophysics at Cambridge University. She was studying at the same university Chandrasekhar had attended, but her interests were in observations rather than theory. Together with her research supervisor Antony Hewish, Bell was observing the effect of the gas and dust of the interstellar medium on the propagation of radio signals from distant astronomical sources. Astronomers had recognised a new class of star-like radio sources and Hewish and Bell were using the interstellar medium to try to investigate the true sizes of the objects. These objects were soon to be known as quasars as we discussed in the last chapter.

Radio waves travelling through the great open spaces between stars occasionally pass through regions of ionised gas called plasma. The presence of this material causes the distant radio objects to scintillate or 'twinkle' when viewed from Earth. This interstellar scintillation is analogous to the effect of the Earth's atmosphere on visible light from stars. We're well aware of a sure way to spot the difference between a planet and a star in the night sky—point-like objects like stars twinkle while planets, that appear larger from Earth, do not. The Cambridge team was attempting to perform a similar size-discrimination exercise on the new radio objects using the interstellar medium instead of the atmosphere. However, in one of the most interesting stories of recent science, they stumbled instead onto a remarkable discovery of a new class of exotic objects—the pulsars.

One of Jocelyn Bell's jobs was to examine the 30 metres of paper tape that recorded the signals from the radio telescope every day. Her task was to identify and chart the positions of fluctuating radio sources. Early in August 1967 she discovered a curious signal on part of the tape. Bell was surprised to see that a similar signal had been recorded earlier in the week from the same part of the sky in the constellation of Vulpecular. Unfortunately the signal only covered about 1 centimetre of tape in both cases and Bell attributed this 'bit of scruff' to local terrestrial radio inter-ference. She put the recordings aside. However, in November a new study required the use of a high-speed recorder. The intriguing signal reappeared and could now be studied with greater resolution. To Bell's amazement, she saw that it consisted of a stream of pulses spaced by a regular 1.3 seconds. She immediately contacted Hewish who was teaching at the time in an undergraduate laboratory in Cambridge. They again decided that the most obvious explanation was an Earth-based source.

Jocelyn Bell was later to say that the terrestrial theory,

> . . . was a very sensible response in the circumstances, but due to a truly remarkable depth of ignorance, I did not see why they could not be from a star. However, he [Hewish] was interested enough to come out to the observatory at transit time the next day, and fortunately . . . the pulses appeared again.

Soon there was no doubt that the signals originated from space. The team was able to eliminate all possible sources of radio noise including car ignitions and observa-tory transmissions. Hewish soon found that the pulses were remarkably periodic and very sharp with pulse durations of only 0.016 seconds. Several weeks of work followed with Hewish and Bell harbouring the nagging possibility that these signals were from some sort of intelligent civilisation deep in space! However, just before Christmas 1967, Bell found another signal in the constellation of Casseopeia which also showed a periodic series of pulses, this time with a 1.2 second spacing. Then she found a third and a fourth source. It was unlikely that four intelligent civilisations were sending similar signals from different parts of the galaxy. The Cambridge team began to feel comfortable that the sources (cheekily named LGM-1 to LGM-4) were associated with astrophysical objects rather than Little Green Men!

What sort of object could be producing such sharp and periodic radio pulses? Astronomers soon realised that the region of space emitting the radio signal would need to be very small. This conclusion came from the sharpness of the pulses, 0.016 seconds in the case of LGM-1. Earlier, in our discussion of quasars, we used the rule in astrophysics that says that a signal can't vary more quickly than the time it takes for light to cross the source region. This puts a limit on the maximum size of the central engines in quasars. In the case of LGM-1, Hewish used Bell's measurements to conclude that the object couldn't be much larger than the Earth. In fact, it could be a lot smaller. Hewish and Bell named the objects pulsars, a shorthand for pulsating stars.

The discovery of pulsars provided justification for Zwicky and Baade's neutron star theory of the early 1930s. Calculations showed that a neutron star, with a mass equal to that of two Suns compacted into a sphere with a radius of 30 kilometres, was quite capable of spinning very fast. In contrast, a white dwarf (four hundred times the radius) would fly apart. Within a year of Bell's discovery, a pulsar was discovered at the centre of the Crab Nebula, with a pulse frequency of 30 times per second! This was no problem for the neutron star theory, but it signalled the end for the white dwarf supporters. The Crab discovery was significant for another reason. It associated, for the first time, a pulsar with the site of a supernova explosion. In other words, it gave weight to the idea that a pulsar was the endpoint of a star's life.

It's now universally accepted that pulsars are spinning neutron stars. That pulsars spin is really no surprise. Many normal stars spin on their axes—our Sun, for example, rotates with a period of about thirty days. A law in physics, known as the Law of Conservation of Angular Momentum, provides the explanation for how the spin rate of a star increases as the star becomes smaller and smaller. In the same way that an ice-skater can quicken a pirouette by pulling her arms closer to her body, a collapsing star will rotate faster as it shrinks. This law can very easily explain how some pulsars can rotate millions of times faster than normal stars.

What about the intense radiation seen coming from these objects? It's believed that the two most important ingredients in the radiation process are this spin rate and the huge magnetic field found around a pulsar. Astronomers have measured magnetic fields with strengths at least a billion times larger than that of our Sun. Theories suggest that when a star collapses into a neutron star, the original magnetic field, embedded as it is in the surface of the star, becomes more concentrated as the star's surface area shrinks. A typical neutron star has a surface area a billion times smaller than our Sun, explaining the increase in the magnetic field by that factor.

A popular model of the pulsar radiation process, the 'dynamo' mechanism, suggests that the rapidly rotating star and magnetic field generate a very large electrical voltage on the surface of the star. The voltage accelerates electrons and protons sitting on the surface to high speeds and the particles are launched into space. Captured by the magnetic field around the star, the particles spiral around the field lines, emitting synchrotron radiation as they go. Most of the radiation is emitted from around the poles of the star where the magnetic field is strongest. The result is similar to two lighthouse beams shining from opposite points on the pulsar. As the star rotates, these polar lighthouse beams may sweep into the view of an observer on Earth, producing the pulses of radiation that give pulsars their name. The wavelength of the emitted radiation depends on the energy of the particles and the strength of the magnetic field—as these parameters increase in magnitude, the radiation becomes shorter in wavelength and more energetic. As we've said, the most common emissions from pulsars are radio waves, but in some young, rapidly rotating pulsars the synchrotron radiation extends from the radio end of the spectrum right through to gamma-rays.

## The ultimate stellar corpses

We've already seen that astronomers believe Active Galactic Nuclei are powered by super-massive black holes. Individual stars can also collapse to form black holes. While it is difficult to be precise, it is thought that stars that begin life ten or twenty times more massive than the Sun will become black holes. This happens at the time of the supernova when the depleted stellar core is so massive that not even neutron degeneracy pressure can halt the catastrophic collapse. Nothing at all can stop the core collapsing to an object of effectively infinite density and zero volume. Physicists describe this state as a 'singularity'.

At a singularity in space all laws of physics become useless. In fact, Einstein's General Theory of Relativity only works until the star's density is equivalent to the mass of the Sun squeezed into the size of a proton! Obviously, with such a huge mass confined to so small a volume, the gravitational field is incredibly strong. Black holes were actually predicted back in 1783 by a Yorkshire astronomer, John Michell, just a few years before the great French mathematician Pierre Simon de Laplace. They realised that a black hole would be produced if the escape velocity from an object exceeded the speed of light. It takes a velocity of 11 kilometres per second (44 000 kilometres per hour) for a rocket to escape the gravitational pull of the Earth. A neutron star with the mass of the Sun would have an escape velocity of 6 per cent of the speed of light, or 20 000 kilometres per second. If that same neutron star was to shrink from a radius of 30 kilometres to 3 kilometres, the escape velocity would exceed the speed of light and the object would become a black hole.

Once a black hole is formed, we can't probe it except through its gravitational effect on surrounding material. No black hole has been positively identified in the Milky Way galaxy, but it's highly likely that one exists in a star system called Cygnus X–1. A strong X-ray source in the Cygnus constellation, it was first discovered in 1964 with rocket-borne detectors. Optical surveys of the same region of space soon identified a large blue star, HD 226868, at the site of X-ray emission. Such a star isn't capable of producing X-rays. However, spectral studies of the star showed that the star was wobbling back and forth with a period of 5.6 days—there appeared to be an invisible partner orbiting around the blue star. From these observations, it's been possible to estimate that the invisible object is nine times the mass of the Sun. Given that a neutron star can't be more massive than about three solar masses, it appears that the object is a black hole.

The X-rays probably come from an accretion disk around the black hole. The disk forms when the intense gravity of the black hole attracts gas from the blue star. As we discussed earlier, the same sort of process goes on in systems containing a neutron star and a normal star in close orbit. Astronomers also believe that much bigger accretion disks are the powerhouses of Active Galactic Nuclei. The model for Cygnus X–1 has been strengthened by observations of the time structure of the X-ray signal. The radiation flickers on and off over periods as short as a thousandth of a second, indicating that the source region is much smaller than a white dwarf and about the size expected for the hot, inner edge of the accretion disk.

We now leave our survey of astrophysics well armed for the discussion in the remainder of the book. We'll see how the field of cosmic ray astrophysics draws from many of the areas that we've already discussed, particularly those on the cutting edge of the bizarre and exotic!

# CHAPTER 5

# COSMIC RAYS IN MODERN SCIENCE

A completely new physical understanding of particle physics came from cosmic ray studies because a completely new energy regime was being explored. This was virgin territory. It is our general experience that science makes breakthroughs when a completely new regime of study becomes accessible. There was a renaissance in astronomy when radio astronomy moved into a completely new part of the electromagnetic spectrum. Optics has undergone a rejuvenation in recent years since a source of coherent light—the laser—has become readily available. Cosmic rays continue to provide information on particles with energies well above those of man-made accelerators but the detailed studies possible using accelerators are not feasible with cosmic rays. The cosmic ray particles arrive at unpredictable times with unpredictable energies and unpredictable composition. Nonetheless, cosmic ray studies still have a record of providing broad information on properties of particle interactions which have later proved correct when detailed accelerator studies became possible.

At the time that Pierre Auger discovered extensive air showers, perhaps the most important aspect of his discovery was that the total energy of the original cosmic ray particle must have been much greater than people had previously dreamed. Even so, those energies are almost a million times less than the ones we will consider later. The greatest long-term legacy of his discovery is that, since the showers spread so much to the side of the original path of the cosmic ray particle, we can use small, simple detectors to record all cosmic rays whose central core (the place where the original particle would have been) is found anywhere within a huge area. All we have to do is have a small number of detectors which are quite widely spaced. Provided that these detectors measure samples of the shower well enough to characterise its main properties (direction, position of its central core), then we are able to deduce most of the information on the original cosmic ray which interests us without covering the ground with many large detectors. For really high energy cosmic rays, we can record all showers falling to the ground within a million or more square metres with only one small 10-square-metre detector. A system, appropriately called the Auger Project, based on just this technique is currently under development.

The complete cascades begun in the air by cosmic ray nuclei have three components. We have already met these in passing. The 'nuclear active' core consists of the remnants of the original particle and its debris (mainly consisting of high energy, strongly interacting particles such as the pions discovered at Bristol). Muons result from the decay of the charged pions in the nuclear active core. They travel in more or less straight lines to the ground with rather little absorption and they can spread to hundreds of metres from the centre of the cascade. The superimposed electromagnetic cascades are begun by the many neutral pions produced along the path of the initiating cosmic ray in the shower core. These immediately decay to give gamma-rays which then multiply in the shower cascade as we shall soon see. The particles in the electromagnetic cascades spread to a typical distance of a few tens of metres from the shower core although we will see that some can reach over a kilometre.

To understand these showers, we need to understand some things about the way in which high energy particles interact and the results of those interactions. The highest energy particles in showers have energies well above those accessible with man-made accelerators and our knowledge of their physics is limited. However, the processes occurring

69

in most shower interactions are at well-studied energies and have been broadly understood for more than fifty years. This means that we can design our detection apparatus and interpret its results with reasonable confidence. There is still much we don't know but cosmic ray physicists have a great deal of practical information about the ways in which the highest energy particles interact.

When they pass through matter such as air or a sheet of lead, all charged particles lose energy as their charge interacts with the electric fields of atoms and molecules in the matter. The most immediate result of this is that loosely bound electrons are knocked out of the atoms and molecules, leaving positively charged ions. This ionisation is what caused Wilson's electroscope to discharge into the surrounding air. Each interaction results in a tiny proportion of the kinetic energy of the original particle being transferred. The process is thus more or less regular and continuous at a loss rate of about 2 MeV per gram per square centimetre. This rather strange unit, a gram per square centimetre, is a measure of the distance travelled by the particle in some material. For instance, in water one of these units would correspond to a distance of one centimetre. (See Appendix 2 for a further description of units used in cosmic ray studies.)

A charged particle cannot avoid losing energy by ionisation. A muon interacts very little with matter apart from by ionisation. As a result, it can travel large distances through the atmosphere and into the ground. However, it is still continuously losing energy by ionisation and to travel through an atmosphere which is 1000 grams per square centimetre thick it must have an initial energy of at least 2000 MeV (at an energy loss rate of 2 MeV per gram per square centimetre). This means that when we detect a muon at a large distance from the centre of a cosmic ray shower, we know from its location that it must have come from a very high altitude (its direction of travel was probably only a degree or so from the original shower direction) and we conclude that its initial energy must have been at least 2 GeV (2000 MeV). If this energy had been lower, the muon would have exhausted its kinetic energy through ionisation loss, slowed down and become unobservable. This simple idea gives us an immediate estimate of the amount of energy involved in a shower. Just allowing 2 GeV for each particle which reaches the ground is a useful first guess at the energy of the primary particle.

Another component of a cosmic ray shower starts from the rapid decay of the uncharged (neutral) pions, unlike the muons which came from the charged pions. This is the electromagnetic cascade (the high energy electrons, positrons and gamma-rays) which interacts in other ways much more readily than the muons. At the highest energies, there are the processes of pair production for gamma-rays and bremsstrahlung for the electrons/positrons. As we saw, these processes have the effect of producing new cascade particles which may themselves then become part of the shower multiplication process.

It works like this. Each gamma-ray from a pion decay produces an electron–positron pair and the electrons (or positrons) lose energy by bremsstrahlung in the form of high energy gamma-rays (photons). You can see that gamma-rays produce electrons which then produce gamma-rays again. This is a cyclic (cascade) process which just increases the number of particles as the average energy of each particle reduces. The energy of the initial particle is progressively shared between increasing numbers of secondary particles and a cascade of particles builds up. Eventually, the energies of the secondary particles become sufficiently low for other processes to become dominant. These new processes do not add further particles to the cascade but still serve to remove energy from it. The cascade then begins to die when its particle energies start to fall below a 'critical' energy of about 80 MeV.

This model contains the essential characteristics of an electromagnetic cascade. In a full atmospheric shower the electromagnetic cascades from individual neutral pions in the core build up and die away. Overall though, the neutral pions are produced progressively in the shower core as it passes through the air. The individual cascades from each of these pions then add to the total number of all shower particles. This number builds and dies rather slowly since it is continually fed by energy from the core. Only a limited proportion of the energy of the initiating particle goes into the cascade particles from each of its interactions.

We can now understand what Wilson and the early workers were studying when they found ionisation in the air. It turns out that there are many more low energy cosmic rays than those with higher energies. Thus there are huge numbers of low energy showers begun at high altitudes in the atmosphere. Their cores of nuclear active particles and resulting electromagnetic cascades rapidly die out. All that is left of these low energy showers at ground level are the muons and they are spread out so much that they reach the ground as independent particles and make up a background of randomly arriving particles at ground level. These 'unaccompanied' muons reach sea level at a rate of about one per square centimetre per minute and constitute about one half of the typical total natural radiation background.

In a sense, for our discussions of cosmic radiation in astrophysics we probably need go no further into particle physics since the necessary ideas are now in place. We should recognise, though, that the particle physics story was to continue with important cosmic ray contributions for at least another decade to the 1950s, before the dominant technique became the use of high energy particle accelerators which provided uncomplicated beams of particles with known energies and compositions. In that decade, a new era was ushered in with the discovery in cosmic rays of a completely new and unexpected group of particles. These were initially known as 'V' particles (later as 'strange' particles) from the characteristic form of their tracks. They had rest masses intermediate between the mesons and the protons. A new mathematical description was needed for them and the number invented by physicists to describe them, their strangeness, reminds us how unexpected they were. However, their study was to lead eventually to our present, apparently almost complete, 'standard' model of particles and their structure.

## Cosmic rays and the solar system

We saw earlier that the magnetic field of the Earth was able to determine which cosmic rays would reach the atmosphere and that the ability to penetrate the magnetic field depended on the rays' energies and their directions. The minimum energy required is a few GeV, much lower than the energy of the cosmic rays which will mainly concern us. At these low energies, incoming cosmic rays are also affected by the magnetic fields found throughout the whole of the solar system. This volume of space which is dominated by the Sun is known as the heliosphere. The fields in the heliosphere have their origin in the Sun and tend to be twisted into a complex pattern by the solar wind of outflowing ionised gas (called a plasma) and the frequent explosive solar flares which send more plasma and high energy particles to distort it. In order to understand these low energy cosmic rays, we would like, ideally, to understand how the heliospheric fields affect the cosmic rays which flow towards us from outside the heliosphere. We might then deduce and understand the properties which the cosmic rays have in interstellar space before they enter the heliosphere. On the other hand, we could guess these properties (or at least their broad features) and use the measured cosmic rays as probes of the properties of the heliosphere. Both types of studies, in fact, occur together and are mutually supportive.

The magnetic field of the Sun is able to bottle up energetic particles for a time. It appears that low energy cosmic ray particles can be accelerated in regions of this field and they can build up in intensity for a time while they are confined. Eventually, the confining magnetic field may break down and liberate large amounts of energy together with the accelerated particles. This is a solar flare and the liberated cosmic ray particles travel out past the Earth into the outer solar system. Cosmic rays produced by the giant solar flare of 23 February 1956 gave us the first direct evidence that magnetic fields originating in the Sun extended at least to five times the distance of the Earth from the Sun into the heliosphere. From direct spacecraft observations we now know that they extend well beyond the orbit of the outermost planet Pluto, to a distance at least one hundred times the distance of the Earth from the Sun. The 1956 flare was so powerful that its effect was measured worldwide, even at the equator which is the hardest region to reach through the Earth's magnetic field. The cosmic ray particles from the Sun first followed a rather direct path to us but, after a few minutes, they were found to come from all directions. The extended solar magnetic field had quickly mixed up their directions. It was realised that the near-relativistic protons were being tossed back and forth by irregularities in that field both inside and well outside the orbit of the Earth.

It is important from the point of view of science to understand our solar system but there is a further vital reason for studying cosmic rays in the heliosphere. Some of the cosmic rays come from the Sun. Admittedly, they have rather low energies (at best up to hundreds of GeV) but this is our one source of cosmic ray particles which can be studied in detail and it should give us clues on how more exotic sources might work. Furthermore, the ways that particles interact with magnetic fields in the heliosphere should be good models of how particles at higher energies are accelerated. For instance, a popular idea of how cosmic rays are produced is that the particles are accelerated in astrophysical shocks (a magnetic equivalent of the shock produced in the air with the passage of a supersonic aircraft). Such shocks occur on a smaller scale in the heliosphere than in exotic distant sources and we can study particles in them either directly with spacecraft or using ground-based detectors. These studies are very revealing since we tend to think of shocks near supernovae or black holes as having rather simple structures. When we study real shocks, we soon find that they have very complex, mixed-up structures.

The study of particles in the heliosphere goes right back to 1930s measurements of cosmic ray intensities using ionisation chambers. Such chambers were routinely operated for many years since it was soon recognised that the intensity of low energy cosmic rays causing ionisation in the atmosphere varied over time, although the reason was unknown. It took many years of data collection to identify the effects being observed. It turned out that there was one effect associated with the 11-year cycle of sunspots on the Sun. This effect caused a variation in cosmic ray intensity of about 1 per cent over that period. One can imagine the difficulty of disentangling this effect with the use of ionisation chambers whose long-term stability could not have been known with confidence at this level. The identification of a slow variation in cosmic ray intensity throughout the sunspot cycle was demonstrated by Scott Forbush in 1954. Forbush has had his name permanently attached to studies of the intensity variation by the naming for him of the short-term, Forbush Decreases associated with sudden and violent solar activity.

In the 1930s, Forbush analysed magnetic data for the US Department of Terrestrial Magnetism but then went on to analyse cosmic ray data from ionisation chambers installed by the department at some of its magnetic observatories. Sudden decreases in ionisation (and hence the amount of cosmic radiation reaching the atmosphere of the Earth) were found to occasionally occur and these could often be correlated with magnetic storms—sudden, short-lived changes in the Earth's magnetic field. These were the Forbush Decreases. In the early 1940s, it was recognised that sometimes the situation was more complex in a surprising way. On some occasions there was an increase in the cosmic ray intensity two or three days before a Forbush Decrease and its magnetic storm. These increases in intensity occurred at the same time as those large explosive events in the solar atmosphere, the solar flares. A model of these processes occurring after a solar flare was developed over the next decade and work continues with direct spacecraft measurements of processes in the heliosphere.

It is sometimes odd to look back and wonder what might have been. In the 1860s, Carrington in London made crucial long-term observations of dark regions on the Sun called sunspots. On one occasion, he saw a strong brightening of a region and also noticed that at the same time there was an effect on his magnetic compass. This was a one-off occasion and he was persuaded that the phenomena may not have been related. In fact he had been the first to observe both a solar flare and its associated magnetic storm.

As we saw, the Aurorae result from cosmic ray particles from the Sun hitting the upper atmosphere of the Earth. Their study has a long and distinguished history, intertwined with the study of cosmic rays. The cosmic rays were known to travel in complex ways through the magnetic field of the Earth but in the 1940s the structure of this field was not understood. Rockets were then flown with instruments to study particles and magnetic fields high in the atmosphere but many phenomena lasted for long periods, much longer than the few minutes accessible with a rocket payload. There was known to be a slow variation in the intensity of cosmic rays with a period which was the same as the solar cycle. It was also known that there could be rapid cosmic ray storms of great intensity and there was a desire to see how these related to solar and geomagnetic information. By the mid-1950s there was a clear need for satellite observations of cosmic rays and the magnetic fields of the Earth. The experiments had been thought out by the cosmic ray physicists and were ready to go. Unfortunately, there was no obvious imperative for urgent satellite flights. The situation changed with the launch of the Soviet satellite Sputnik. This was at the height of the Cold War and the ability to launch satellites quickly became an important propaganda issue for both sides.

When a solar flare occurs, many high energy particles which were trapped in a part of the solar magnetic field are suddenly liberated and travel directly outwards from the Sun. These are the particles which produce a prompt cosmic ray increase. The Sun is continually losing gas from its outer atmosphere as a 'solar wind', predicted by the American Eugene Parker in the 1950s. This gas is ionised and has a high electrical conductivity. Such a state of matter is called a plasma and plasmas, as a consequence of their high conductivity, carry magnetic fields with them. Those fields are said to be 'frozen in'. The magnetic field around the Sun thus has a structure which is dynamic and is associated with the continuous outflow of the solar wind. If a solar flare occurs, a great deal of energy in the form of charged particles is added to the solar wind and this extra plasma travels faster than the main body of the wind. A shock structure is set up in the solar wind and this makes the continuous flow of cosmic ray particles into the heliosphere become disrupted. The results are that the shock structure in the wind and its associated magnetic field affect the field of the Earth and cause a magnetic storm. Additionally, the number of cosmic rays (which originally may have increased due to direct particles from the Sun) now reduces as these are mainly from outside the heliosphere and their incoming flow is retarded by the outflowing energy of the flare. This is the explanation of the series of cosmic ray effects we can find following a solar flare.

The study of these shocks by spacecraft gives us our most direct information on shocks in the Cosmos and we know that they accelerate particles. Spacecraft are important but they are expensive. Fortunately, many of these studies can be carried out by Earth-based detectors. Ground-based 'neutron monitors' which are sensitive to the lowest energy cosmic rays have provided us, for forty years or more, with invaluable information. These data now cover two full solar cycles but the long-term monitoring role of the detectors is becoming even more valuable as the processes in the Sun and its surrounding heliosphere are recognised to be complex and to directly affect life and commerce on the Earth.

We can regard the solar wind as a form of weather which particularly affects the outer regions of the Earth's environment. It is harsh and inhospitable. Charged particles can be directly dangerous to human life and magnetic fields can be dangerous to the machines we need for life. The particle radiation in space from a solar flare (or even the background cosmic radiation from outside the solar system) can kill. At the surface of the Earth we are protected from much of the radiation by our magnetic field and by an absorbing atmosphere. Yet there are proposals to occupy space stations for extended periods and our commercial aircraft are flying higher than ever with less covering atmosphere for protection. This results in increased radiation exposure from cosmic rays for aircraft crews and passengers and increased concern for astronauts and their equipment, particularly at times of increased solar activity. Furthermore, magnetic storms produced by the solar weather can induce electrical surges in power lines. As the sizes of power grids have increased, the damage induced by these surges is potentially increasing and damage measured in hundreds of millions of dollars has already occurred. There is a pressing need to improve our knowledge of these processes to enable us to give warning to those who risk radiation exposure and to instrumentalities whose businesses can be ruined.

## Carbon–14 dating

Low energy cosmic ray protons can efficiently produce neutrons in the atmosphere. These, being uncharged, are able to penetrate the atmosphere and can be detected at the ground with neutron monitors. The neutrons not only provide us with a direct detection technique for low energy cosmic rays but they are continually bombarding all the material in the air and on the ground. In particular, they bombard atmospheric carbon nuclei (in carbon dioxide), converting some of those nuclei from the normal carbon–12 isotope to the isotope carbon–14. Living plants absorb this carbon–14 with other carbon isotopes as part of their energy cycle. The proportion of carbon–14 in the overall atmospheric carbon and in living systems such as plants is kept roughly constant at a known value by the roughly constant cosmic ray bombardment.

When a plant dies, carbon ceases to be cycled into the atmosphere and the balance between the various isotopes is lost. In particular, the carbon–14 nuclei are radioactive and decay at a predictable rate. As a consequence, any dead plant material will progressively reduce its proportion of carbon–14 and a measurement of that proportion will give the time elapsed since the plant died. Half of the carbon–14 decays away in 5760 years and with modern techniques for determining the abundances of the various carbon isotopes in even small samples, this provides a powerful technique for dating artefacts from recent millennia.

Notice that we had to assume a constant intensity of cosmic rays in order to use this technique with confidence. If we know the real age of some artefacts in an independent way, we can check that constancy. This reveals that cosmic rays (which are controlled in intensity by the activity of the Sun and the magnetic field of the Earth) have arrived at an almost constant rate over the past few thousand years.

## Cosmic ray detectors

We have seen that cosmic radiation was discovered through observations of the ionisation produced by its energetic charged particles. The early ionisation detectors have little use now and have been superseded by other techniques for observing high energy charged particles quickly and efficiently.

There is always a background of unwanted radiation which can mask the effects of the cosmic ray particles. As a result, it is usually necessary to have detecting equipment which responds quickly so that there is no confusion between observing the wanted effect of the cosmic rays and the effect of 'noise' contamination. As we've seen, the first major development after the ionisation chamber (which dominated before the 1930s) was the Geiger–Müller (or just Geiger) counter. This works by responding to ionisation produced by a cosmic ray as it passes through the gas contained by the counter. A strong electric field is set up within the counter so that ions produced in the gas by the passage of the cosmic ray particle are accelerated sufficiently to produce further ions when they hit other gas molecules. In this way, a charged cascade is set up and a large electronic signal builds up as the whole counter rapidly discharges in a relatively uncontrolled way, rather like a spark moving through the gas. This means that a simple counter can be built which produces a large electrical signal. It was an ideal way of detecting single charged particles given the insensitive electronic apparatus which was available in the early decades of the twentieth century. On the other hand, the counter was slow to recover from a discharge and could not cope with a high rate of particles. Also, it could only indicate that a particle had passed through and did not give an indication of the initial amount of ionisation (a good clue to the type of cosmic ray particle or even how many particles there were). A better detector was the proportional counter which is similar in overall concept but is operated under more controlled conditions so that the size of the resulting signal is proportional to the amount of ionisation produced in the counter by the passage of the particle.

Arrays of Geiger or proportional counters were built when large detectors were needed but a more simple, and often more attractive, alternative became available with the advent of the scintillation detector. This detector depends on two components. The first is a material which ˬemits light (scintillates) when an energetic charged particle passes through it. Such materials had been used from early times in the study of particles by Rutherford and his colleagues who used microscopes to view a thin film of zinc sulphide to detect radiation. The passage of a particle showed up as a flash of light which was counted by the observer. The process was very tedious and also inefficient since the zinc sulphide was opaque and only light produced near the surface could be detected. The technique was also very dependent on the alertness and eyesight of the observer. Eventually, new materials were found to scintillate. Certain plastics and liquids could then be made into cheap large-area detectors. Also, certain crystals, though much more expensive, could be made into detectors whose light output was accurately proportional to the energies of the electrons and gamma-rays entering the detector. The second component of the scintillation detector is something to record the light. This is almost invariably a photomultiplier, a photo-electric cell which responds with great speed and sensitivity to the burst of light from the scintillating material. Plastic scintillators can be used directly in space as components of detectors aboard spacecraft to measure the lower energy cosmic rays. Crystal scintillators have also often been used in space as the fundamental part of X-ray and gamma-ray telescopes because of their ability to completely absorb those photons and to produce a quantity of light proportional to the photon energy. Used in this way they are known as spectrometers.

Plastic scintillators come into their own in the ground-level detection and measurement of extensive air showers. These showers often contain huge numbers of particles but they are spread over large areas and so the number of particles per square metre on the ground can be quite small. Economical observation of such showers demands the use of cheap, large-area detectors and the plastic scintillator is often the detector of choice for this purpose. One-square-metre detectors are very common and detectors with areas up to ten times this can be built in single units. An air shower detector array usually consists of many of these units (perhaps up to one thousand) widely spaced and operated 'in coincidence' so that the recording of data is initiated when a number of the detectors register the passage of particles at the same time. The number of particles passing through each detector is then measured electronically as is the time of passage of the particles to a precision of perhaps one billionth of a second.

To understand the purpose of this 'fast timing', we need to look again at the structure of the air shower cascade. As we have seen, the particles in an air shower all have very high energies, typically much more than the energies associated with their rest masses. For instance, the energy of a typical electron in the air shower cascade might be 40 MeV (as we might expect, somewhere below the critical energy of 80 MeV) but the rest mass of the electron is only 0.5 MeV. In the jargon of physics, these particles are highly relativistic (see Appendix 1) and so travel at speeds which, for practical purposes, are indistinguishable from the speed of light. This means that everything in the shower travels at the same speed and all the shower constituents will remain together as a bunch through the atmosphere. This is not quite true since Coulomb scattering of the electrons and the divergence of the muon directions cause the shower to spread to the side into a saucer-like disk. The disk (travelling at close to the speed of light) is a couple of metres thick at the centre, increasing to several metres at a distance of hundreds of metres laterally from its core. It also curves back slightly towards its outer edges.

For many practical purposes, the shower disk can be thought of as a thin, flat, circular sheet travelling at 300 000 000 metres per second (the speed of light). As it reaches the array of detectors, its particles will pass through them in an order and at relative times which will depend on the arrival direction of the air shower. A vertical shower will arrive at all detectors at the same time but a shower from another direction will progressively sweep across the array. The array's fast-timing electronics record these times and the direction of the shower can then be deduced mathematically. With shower dimensions typically a hundred metres or so and electronic timing good to a few nanoseconds, the direction can usually be reconstructed with an uncertainty of about 1 degree.

There is a rather strange effect which can be utilised in the detection of relativistic charged particles and this is named after its discoverer (in the 1920s), Pavel Cerenkov. The speed of light is the ultimate limit of speed for any particle. This is the speed at which light travels through the vacuum of space. When light is in a transparent material, it moves more slowly than this. The amount of slowing is described by a number called the refractive index of the material. The refractive index is usually a number between 1 and 2 and the local speed travelled by light is given by the speed for free space divided by this number.

So a piece of glass with a refractive index of 1.5 will carry light at a speed of 300 000/1.5 (or 200 000) kilometres per second. This slowing results from the way in which the molecules in the material interact with the light beam and does not apply to relativistic particles travelling through the transparent material. So relativistic particles can actually travel faster than the local speed of light in transparent materials. This strange situation, apparently precluded by the Special Theory of Relativity, had been the subject of much speculation even before the time of Einstein. It has an equally strange consequence.

We are familiar with at least two other situations in which a *source* of energy travels faster than the speed at which the energy itself can travel. These are when an aircraft travels faster than the speed of sound and when a boat travels faster than water waves. In both these cases, a strong (shock) wave is formed by the source—the aircraft or the boat. A shock wave is also formed when a charged particle travels faster than the local speed of light but this is now a shock of electromagnetic energy which results in the emission of light rather than sound or water wave energy. This light is known as Cerenkov light.

Cerenkov light is emitted in a cone around the direction of motion of the particle. In water or glass this cone angle is around 40 degrees. In a gas such as air Cerenkov emission still occurs, but with a very small cone angle since the refractive index is very close to 1. Water and glass are quite efficient producers of Cerenkov light since they have large refractive indices. It turns out that the quantity of light which is produced increases with the value of the refractive index. Cerenkov emitters can be used rather like scintillators in charged-particle detectors when combined with a photomultiplier to detect the light.

The use of Cerenkov light developed into a powerful technique in the 1950s as sensitive, fast-response photo-multiplier light detectors became available. These could record flashes of light from individual particles. A particu-larly interesting Cerenkov detector for cosmic ray studies is the deep-water detector pioneered in London and then employed with great success in the Haverah Park air shower array in Yorkshire, England. These detectors consisted of large, closed water tanks with a depth of 1.2 metres. The water was viewed by photomultipliers dipped into its surface from above. The electromagnetic (sometimes called 'soft' because it is less penetrating than the 'hard' muons and nuclear particles) component produces light mainly in the top third of the water when an air shower passes but the whole of the tank was sensitive to the more penetrating muons. This produced a detector signal which was a combination of the signals from the electromagnetic and muon components.

As we saw, high energy particles can also produce Cerenkov radiation in the atmosphere. The refractive index is close to 1 (about 1.00027 at ground level) but, providing that the particles are energetic (above about 20 MeV for electrons), Cerenkov light can be produced by many of the particles in an air shower. The light is quite weak (again because the refractive index is close to 1) but there can be many particles in a shower and so, often with large mirrors to collect the light, such emission is readily detectable on clear, moonless nights. Such atmospheric Cerenkov studies make two important types of observation possible.

First, since the light is produced by the air shower particles all the way through the atmosphere, it is possible to use measurements of the light to infer the way in which a shower developed in the atmosphere. This turns out to be a vital tool in deducing the mass of the primary cosmic ray particle for showers with energies of about one million GeV. Second, low energy showers may have very few particles which actually reach the ground but they may still be detected through their atmospheric Cerenkov light, provided that large mirrors are used. This technique is used with great success to study the showers produced by cosmic gamma-rays with energies slightly above the maximum energies practically observable using satellites. (Very energetic gamma-rays can initiate showers just like cosmic rays.) Again, properties of the Cerenkov light itself can be used to obtain information on the nature of the primary particle and, in this case, the information can be used to reduce the number of non-gamma-ray initiated showers which are accepted since these measurements are made by astronomers wishing to study only the gamma-rays.

In the 1960s, a possible alternative technique was suggested which came to spectacularly successful fruition in the 1980s. This technique involves detecting the fluorescent light produced by shower particles traversing the atmospheric gas. We saw that the shower particles lose energy by transferring energy to the atmosphere by ionisation. This involves the removal of an electron from an atom and requires the expenditure of a substantial amount of energy. For this to happen, a shower particle must pass very close to the atmospheric molecule to ensure that enough energy can be transferred. Most molecules are too far away for such a large energy transfer to be successful and they only have a feeble force acting on them. This feeble force can still affect the molecule and change the distribution of energies within it. The redistributed arrangement of electrons is usually unstable and the molecule will eventually return to its original arrangement and give up the energy transferred to it. This energy is usually given up in the form of a photon of light with quite a specific wavelength and the emission process is called fluorescence.

Atmospheric nitrogen produces blue fluorescent light in this way and such blue light is ideally suited for detection by photomultipliers which can be used for its observation when there is no moon and no cloud. The fluorescence process produces less light towards the detector than Cerenkov emission but optical filters can be used to select the exact light wavelengths of interest and the technique can be applied to studying the very rare highest energy cosmic rays. The great strength of the technique is that, unlike Cerenkov light, the fluorescent light is emitted in all directions so that showers can be detected side-on as well as when they travel directly towards the detecting equipment. This means that all showers may be readily detected making the technique ideal for survey work where a complete set of shower detections is needed. Atmospheric fluorescent light is fundamental to the operation of the Fly's Eye cosmic ray detector and its successors, HiRes and the proposed mammoth Auger array, which we will discuss in some detail later.

# CHAPTER 6

# PROPERTIES OF THE PRIMARY COSMIC RAYS

A lot is now known about the cosmic rays which reach the Earth and we will need this information as we search to understand how and where Nature produced them. In this chapter, we will look at those properties of cosmic rays which have been well measured and are now generally quite well accepted. These are properties such as the composition of the primary cosmic ray particles (are they protons—nuclei of hydrogen atoms, oxygen nuclei, iron nuclei or whatever?), their energies and the relative numbers of cosmic rays with different energies, and their direction or the ways in which they have travelled to us. Such properties are all difficult to measure since it is necessary to make measurements on single particles coming at unpredictable times from unpredicted directions. The result is that we often know only broad properties of the particles. We tend to know these properties best at low cosmic ray energies where the particles are plentiful and can be measured directly with detectors in space.

Cosmic ray directions are affected by magnetic fields. At lower cosmic ray energies, the fields of the Earth and the heliosphere are important. At energies above about $10^{11}$eV, we are mainly concerned with those fields found in our Milky Way galaxy. The cosmic rays are deflected less as their energy increases and they follow great looping spiral paths, the scale of which increases in proportion to the cosmic ray energy and inversely with the charge on the particle. This scale begins to be comparable with the size

**Figure 6.1 The cosmic ray energy spectrum**

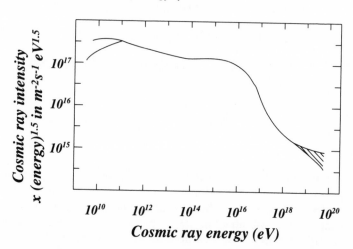

This is the cosmic ray spectrum in Figure 1.1 redrawn to emphasise the features of the spectrum, the 'knee' at about $10^{16}eV$ and the 'ankle' at about $10^{19}eV$. We have multiplied the intensity scale by the cosmic ray energy to the power of 1.5. The divergence at the left is due to variations caused by the solar cycle. The divergence at the right is due to a lack of data at those energies.

of our galaxy at cosmic ray energies of over $10^{15}eV$. Below such energies we can be reasonably confident that the particles wander more or less randomly for very long times within the Milky Way. We tend to think of these as being trapped within the galaxy for those times. It would be helpful to know for how long and, as we will see, these lifetimes are measurable. At higher energies, we assume that the cosmic rays leave our galaxy quite quickly and their directions are not so mixed up as at lower energies.

We have seen that cosmic ray energies vary over a huge range. The ratio of the lowest to the highest cosmic ray energy is roughly the same as one millimetre to the size of the Earth (it is also about as big as the ratio of wavelengths at the extreme ends of the spectrum studied by astronomers—from radio astronomy right through to gamma-ray astronomy). Also notice that the 'rate of particles' scale at the left of Figure 1.1 changes by many times more, something like the ratio of one millimetre to the size of the visible Universe. On the whole, the spectrum does not change its structure much over the full span of the range.

At the lowest energies, the spectrum does seem to change abruptly but that is because we are not able to get a clear view of the cosmic rays from here on Earth. At energies below about $10^9$eV, cosmic ray particles have difficulty in approaching us through the magnetised solar wind which constantly flows outwards from the Sun. We cannot then directly measure the spectrum in interstellar space from such observations although at times, when the Sun is least active, we can see somewhat lower energies. We say that the spectrum which is shown has been modulated, or changed, from the true spectrum to be found in the galaxy between stars.

As an aside, this spectrum can be used to calculate the energy density of cosmic rays. This is a quantity which is important to our theoretical understanding of cosmic rays. It tells us on average how much energy there is for each unit of volume in whatever part of the Universe we are studying. For instance, if we consider all cosmic rays with energies above $10^9$eV in our galaxy (the lower limit we can directly observe), the energy density is about 1 electron volt per cubic centimetre. It is either a remarkable coincidence or a clue yet to be understood that this value is almost exactly the same as the energy density of starlight in our galaxy and the energy density of the galactic magnetic field. We sometimes say that there is equipartition of energy. There may be a good reason for this since the cosmic rays are continuously sharing their energy with the magnetic field as they bounce off high field regions. The other strange thing about this number is that, in a sense, it is artificial. It just happens that we do not know in detail what happens to the spectrum at lower energies. We might try to guess what could happen at lower, unobserved, energies. If the spectrum just tapers off as we go lower, then our last estimate was probably about right.

However, it is fascinating to speculate on what happens if we just extend the spectrum back without a change to the straight line in Figure 1.1. In this case we need to recognise two forms of energy. The kinetic energy density goes up quite a bit in this case but not spectacularly. It is interesting but not awesome. However, it becomes impressive if we remember that mass has energy too (see Appendix 1). A proton has a rest mass of $10^9$eV and so every single lower energy particle carries that equivalent amount of energy. With such a steep spectrum, the energy (or mass) density is soon huge as we extend the spectrum progressively down in energy. Studies of the way our galaxy rotates tell us that it contains much more mass than we have been able to identify observationally. Our speculation about the spectrum of the low energy cosmic rays suggests that perhaps here we can find enough matter to explain the way our galaxy rotates. If we push our speculation to rather lower energies, we have enough energy to affect the evolution of the Universe. We could call this a hot, dark matter scenario. We need to find at what stage, if any, this suggestion fails.

A careful examination of the cosmic ray energy spectrum reveals that at intermediate energies of about $10^{15}$eV the spectrum begins to steepen somewhat and above about $10^{18}$eV it flattens again. These features are emphasised when the overall steepness of the spectrum is artificially removed. It can then be seen why they are referred to as the knee and ankle.

At energies just above about $10^9$eV, cosmic rays are unable to penetrate far into our atmosphere and cannot be studied from the ground. However, they are plentiful and satellites are ideal for their study. Furthermore, this is a range of energies for which techniques of nuclear physics are well developed and can be applied to make direct observations of these particles in space. From the earliest times of the space age, such measurements have been high on the list of jobs for spacecraft. The detailed ways that a particle detector responds to a particle passing through it depend on the physical characteristics of the particle itself due to its composition (its rest mass and charge) and its energy. If the particle passes through two or more different types of detector, it is possible to separate the effects of composition and energy to identify both of these qualities of the particle unambiguously. This can be done in space at modest energies and the composition of individual cosmic rays identified even to the extent of identifying which isotope of an element has been detected. At rather higher energies, this type of measurement becomes more difficult as the particles become more penetrating and the number of available particles drops dramatically.

It is natural to examine the types of nuclei present in the low energy cosmic rays and to ask whether these are what we would expect. These are all nuclei like those at the centres of familiar atoms: hydrogen, helium, carbon etc. On the whole, they are found in rather similar proportions (called abundances) to those which we find in our solar system. However, there are some glaring discrepancies which give us insight into the origin of these particles.

The clearest discrepancy is found for the elements with atomic numbers 3, 4 and 5 (lithium, beryllium and boron). In the table of elements, these are the ones near the start (immediately after hydrogen and helium). For this reason, they tend to get lumped together as 'light' nuclei. In the Universe as we know it, these light nuclei are hardly to be found. They are so rare that there is only about one of each of these nuclei for a billion or more of hydrogen or helium. When we look at the cosmic rays, we find that the light nuclei are clearly present and in numbers which are perhaps one hundred thousand times greater than we would have expected. Such a result seems strange since the elements immediately past these in the table of elements have more or less the right abundances. At first sight it seems to say that, for some strange reason, the sources of cosmic rays are made up of more or less the same material as the rest of the Universe except that there are lots more lithium, beryllium and boron elements. This is unconvincing and we would like to find an alternative source for these extra elements.

The likely answer would seem to be that most of the light nuclei do not come directly from the cosmic ray sources at all. We know that the cosmic rays travel long and tortuous routes through our Milky Way galaxy from their sources to us. We also know that our galaxy, while apparently being made up largely of stars, has tenuous gas between the stars. This gas is probably better described as a very good vacuum with only about one atom per cubic centimetre but, to the cosmic ray particles which travel for aeons through the interstellar medium, it acts as a target for nuclear interactions. When energetic nuclei hit targets made up of other nuclei, it is likely that both will break up into smaller fragments. This is certainly the case for cosmic rays with energies around $10^{10}$eV since the forces holding nuclei together have energies measured in millions of electron volts, over a thousand times less than the energy of the cosmic rays. The rather abundant heavier nuclei such as carbon, oxygen and nitrogen are up to twice the mass of the overabundant light nuclei and when they break up they have a natural tendency to fragment into debris consisting of the light nuclei and other material. We thus regard those light nuclei as secondary and having been created in the passage of other cosmic rays between their source and us. This process is known as spallation.

If we can understand the spallation process properly we can make an estimate of how far, on an average, the cosmic rays have travelled from their source. We need also to know something about how many target atoms there are in the interstellar gas and how likely the nuclear reactions are. There is a problem which we must recognise. We have to assume in this that the interstellar gas is pretty uniform in the parts of the galaxy traversed by the cosmic rays. This is probably not a very good assumption since we know that the regions in our galaxy with high magnetic fields are related to the regions with higher densities of gas and the paths of the cosmic rays are determined by those very same magnetic fields. It is not easy to untangle these related effects so we first ask how much material has been traversed. This amounts to the equivalent of a few centimetres of water. It is not much but it corresponds to a great distance (or time travelled) if the density of the interstellar medium is measured in atoms per cubic centimetre. In this way, we find a value of a few million years for the lifetime of a typical cosmic ray particle (with an energy of a few times $10^9$eV) in our galaxy.

It turns out that there is a way of checking this lifetime value and this again is possible because of the detailed information we have on the composition of these quite low energy cosmic rays. Beryllium–10 is a radioactive nucleus with a half-life of 1.6 million years. This means that, if one originally had a certain number of beryllium–10 nuclei, then after 1.6 million years only one half would remain. The other half would have spontaneously split into other nuclei. This decay process is unpredictable for any specific nucleus but averages out when many nuclei are studied. After a further 1.6 million years the remaining nuclei would again have halved in number. The result is that we can measure the number of beryllium–10 nuclei in the observed cosmic rays, estimate how many there would have been without any decay process (this must take into account the production process of beryllium just discussed) and then deduce the time required for the appropriate numbers of nuclei to decay. The value of this time is a few times greater than the value we got previously from a consideration of the abundance ratios. We presume then that the cosmic ray nuclei have actually travelled for that extended time but that much of the time has been spent in places where there are rather few target nuclei for spallation. So the cosmic rays appear to have been excluded by some process from where most of the gas exists in the galaxy or, perhaps more likely, they have spent time outside the disk of our galaxy where there are still magnetic fields but the density of gas atoms is low.

Our picture of the passage of cosmic rays through our galaxy was based on cosmic rays of a particular energy—a few GeV. We find that the results change progressively as the cosmic ray energies increase. That is, when we ask what thickness of absorber has been traversed, we get values which drop as the energies increase. It is as though the Milky Way galaxy is becoming less good at holding in the higher energy cosmic rays. This idea of the Milky Way as a 'leaky box' on the whole works quite well.

It would be good to have a way of studying the distribution of cosmic rays throughout our galaxy. There are observations which make this possible although they are difficult to make and their interpretation depends on a good deal of computer modelling. We know that cosmic ray particles interact with the diffuse gas throughout our galaxy. One result of many of these interactions is that gamma-rays are produced. The gamma-rays then travel in straight lines and some can be detected by instruments on spacecraft such as NASA's Compton Gamma Ray Observatory (CGRO) and its predecessor, COS B. Not surprisingly, the distribution of these gamma-rays over the sky shows the general structure of the Milky Way, confirming our general ideas of their origin.

In order for the gamma-rays to be produced, one must have both galactic gas and cosmic rays. The computer models use the best information we have on the distribution of gas throughout the galaxy and then make various assumptions (or models) of the distribution of cosmic rays to get a fit with the observed directions of the gamma-rays. It turns out that the cosmic rays, in fact, do gradually reduce in intensity with increasing distance from the central regions of our galaxy and so we are probably on firm ground in believing that most of the low energy cosmic rays have their origins in galactic objects.

We have reached the conclusion that the lower and medium energy cosmic rays have properties which suggest that their origins are within the main body of our Milky Way galaxy. The situation changes as energies increase further and particles remain for less time in the vicinity of the galactic plane. What actually changes in detail as the energy increases?

94

Measurements of the cosmic ray mass get progressively more difficult as the energies increase. Results from high altitude balloon experiments by the Japanese–American Cooperative Emulsion Experiment (JACEE) collaboration are now giving us vital and intriguing data up to energies approaching $10^{15}$eV. These experiments are based on observations from high altitude balloons, floating within 0.5 per cent of the top of the atmosphere. The result is that cosmic rays can be observed with very little interference from the atmospheric gas although corrections for atmospheric effects still have to be made. The JACEE experiments are carried out by a collaboration of Japanese, US and Polish researchers and data have been accumulated over a long series of balloon flights. The basic experiment consists of detecting cosmic rays, measuring their charges and then measuring their energies. These measurements are particularly important because the range of energies which can be studied is large. This is because the detection system is quite large and, with re-use of the detector over many flights, its observing time is long. The use of an experiment with a large product of area and time means that a reasonable number of energetic cosmic rays is detected even though they are rare.

The JACEE experiment is now clearly showing that, as the energy increases, the proportion of heavier nuclei (those further along the series of elements) increases. If we extend the trend of these results, we get the intriguing suggestion that at energies near the knee of the cosmic ray energy spectrum (in the vicinity of an energy of a few times $10^{15}$eV), heavy nuclei may be the dominant component. This is totally contrary to the overall abundances in the Universe which are dominated by hydrogen and helium.

Is it possible to test this suggestion for the composition at energies near $10^{15}$eV? Certainly it is in principle but the experimental problems are daunting. At the present time, attempts have been made but there is no clear-cut answer. The difficulties are related to the dramatic reduction in the number of particles as the energy increases and to their increased penetrating power. At an energy of about $10^{15}$eV, cosmic rays arrive at a rate of only about one hundred per square metre per year. This is clearly bad for balloon or satellite experiments which have, at best, areas of only a few square metres. Additionally, the determination of the particle energy usually depends on absorbing most of the energy in the detector and, at these energies, this requires a kilogram of absorber for each square centimetre of detector area. A 1-square-metre detector would require a mass of 10 tonnes and this is very expensive to fly in a satellite or a balloon payload (although this was achieved in an early Soviet experiment). The best remaining approach we know is to study the extensive air shower cascades produced by these particles and attempt to determine the mass of the primary particle and the cascade energy in an indirect way.

Determining the energy for a primary cosmic ray is possible through measurements which tell us how many particles there are in the shower produced by the particle. The mass determination is more difficult since it is really an attempt to measure the mass of a single nucleus at a distance of perhaps 10 kilometres. The technique usually employed is to study the detailed development of the shower in the atmosphere and to find, particularly, the atmospheric depth at which the number of particles in the shower reaches its maximum value. This is a delicate and difficult measurement which, ominously, depends on our theoretical understanding of particle physics at these energies and is interconnected with the energy determination. When experimental measurements are made of two quantities and those quantities are not independent then there are likely to be many subtle pitfalls.

The general idea is that massive nuclei are larger (in physics jargon, they have a larger cross section) than light nuclei (or protons) and so interact earlier in their passage through the atmospheric gas. Additionally, as the heavier nuclei break up, the resulting cascades develop more quickly. These effects combine to give the result that heavy nucleus primaries should produce cascades which reach their maximum development at substantially higher altitudes than those initiated by protons or light nuclei. This difference could be by as much as one-tenth of the total atmospheric thickness. One might expect that this would be easy to discern. Not so.

It turns out that at energies of about $10^{15}$eV the particle cascades certainly contain many particles (a few hundred thousand) but the measurement requirements are too subtle to select with certainty between different models of the composition. A number of possible measurements can be made. A method which was popular when the problem first became important was to measure the ratio of the number of muons in the shower to the number of other particles (mainly the electromagnetic component) when the shower reached the ground. The muon numbers are not particularly dependent on the height of shower maximum but, since the electromagnetic component is rapidly attenuating, a shower which develops early will have attenuated that component a great deal by the time it reaches the ground. The result should be that a shower initiated by an iron nucleus will have a greater ratio of muon numbers to electromagnetic particles than a late-developing proton shower. The principle is fine but usually there are not enough muons. There are typically one-tenth as many muons as other particles in a shower at these energies and these are spread over an area of perhaps 100 000 square metres. Given that there might be only 10 000 muons, they will be few and far between. Many large and expensive detectors are required for such an experiment.

An alternative, and potentially more attractive, technique is to directly measure the height of shower maximum above the observer using the forward-directed Cerenkov light generated by the shower. This can be done either by measuring the spread of the light on the ground (light from high up spreads more than light from lower down to a first approximation) or by measuring the spread in time of the light pulse (for less obvious reasons a large time spread results from a late-developing air shower). These techniques again are fine in theory, but their detailed interpretation causes great problems.

A fundamental difficulty with all these air shower techniques results from the fact that a pair of showers initiated by two identical particles of identical energy will be very unlikely to develop in exactly the same way. On an average we would expect the developments to be the same but showers develop through a number of statistically determined processes and they are subject to great fluctuations in their development. One has only to think about the first interaction to see this. Suppose that, on average, a proton would interact after travelling through one-tenth of the atmosphere and a heavy nucleus after one-thirtieth. The difference is great. On the other hand, a particular proton could easily, by chance, happen to interact with an air nucleus near the top of the atmosphere—even earlier than an average heavy nucleus. Alternatively, the heavy nucleus stands a reasonable chance of travelling through three times its average distance before interacting—in this case acting very much like a proton. This means that in our measurements we have to be acutely aware of subtle effects associated with these fluctuations in development. It turns out that they are very difficult to cope with.

It may seem technically trivial, but a deep problem is associated with simply detecting the shower in the first place. An air shower array responds to the particles which reach the ground. Because of its reduced attenuation, we have just seen that a shower which develops deep in the atmosphere will contain more particles at ground level than we would have expected on average. One which develops early will have correspondingly fewer particles. Imagine a mixture of particle types arriving at the top of the atmosphere to be detected by a system which responds to the number of particles at ground level. The light nuclei will, on average, interact one-tenth of the way through the atmosphere but could wait until three times that distance, at around a third of the total atmospheric depth. In this case (not possible for heavy nuclei), for the same number of particles at shower maximum (that is, the same energy) the light nucleus shower would have five to ten times as many particles at sea level and would be much more likely to be detected.

Even worse, two protons with the same energy would either be detected or not depending on whether, by chance, the development of the shower for one fluctuated lower in the atmosphere than the other or vice versa. The resulting effect is called trigger bias in which the selection of showers by the air shower array results in a complicated and biased mixture of showers being recorded. In the case of an equal mixture of protons and iron nuclei in the primary cosmic rays, most of the lower energy showers detected by the array would actually be due to protons with an average depth of shower maximum much nearer to the ground than their true average. These are reasons why the composition near the knee of the spectrum is still the subject of controversy although many measurement attempts have been made at medium energies.

## Why composition might be as we find it

Experiments at the lower energies suggest heavy nuclei become dominant at the knee of the energy spectrum even though we cannot presently confirm this with certainty. If this conjecture is correct, would that be a reasonable result? On the face of it, the answer is yes although there are perplexing details which are still to be placed convincingly into the picture. The idea is to explain the steepening of the spectrum and change in the composition by a single concept or model.

Our galaxy is permeated by magnetic fields which constrain the charged cosmic ray particles to move only slowly from their sources within the galaxy to its edge where they are lost to intergalactic space. This process is one in which the cosmic rays follow spiral paths in those fields which are mainly in the spiral arms of the galaxy. The magnetic fields, like the spiral arms, have a general structure which is circular about the centre of the galaxy. The field may well follow the spiral arms. However, the field structure is grossly distorted by many twists and turns as the field has been distorted over the millennia by supernovae and the rotation of the galaxy. As a result, although the field lines themselves are continuous and on average neither lead into nor out of our galaxy, the cosmic rays can slowly move from one line to another, particularly when there are sharp twists and turns. Inevitably, the cosmic rays gradually move to outer galactic regions and are lost.

The rate at which this process happens depends on how many sharp twists in the magnetic fields the cosmic rays come across. But what do we mean by a sharp twist? Roughly speaking, one might imagine this as one which twists around in a distance less than the distance it takes the cosmic ray to complete one cycle of its spiral path (called the Larmor radius). In this case, the cosmic ray will lose its attachment to a particular path in the magnetic field and will wander to another. The Larmor radius increases with the energy of the cosmic ray particle so that high energy particles have paths with large radii and will encounter many sharp twists. As a result, the energetic particles tend to leave the galaxy faster than the lower energy ones and the total number of high energy cosmic rays in the galaxy at any time is reduced compared to lower energy rays. This model is used to argue for a knee in the energy spectrum above which the number of high energy cosmic rays is less than it would be if the low energy spectrum had been indefinitely extended.

What about the charge or composition of the cosmic rays? Particles with low charge (the lighter nuclei) have large Larmor radii as they move through the magnetic fields of the galaxy. The radii are proportional to the inverse of the charge—an iron nucleus will have a Larmor radius which is only one-twenty-sixth of that of a proton with the same energy. Following our previous argument, the proton cosmic rays will be lost from the galaxy at lower energies than the heavier nuclei such as iron. A result will be that, from well below the knee, there will be a progressive reduction in the number of light nuclei in the cosmic rays until, at the knee, all of the light ones will have been lost. Thus, below the knee there will be a progressive change in the observed composition as the cosmic ray energy increases to almost pure heavy nuclei at the knee—exactly as the JACEE experiment seemed to suggest.

Notice that the cosmic ray composition at the source, whatever that is, is not required to change. It is just that the cosmic rays in the galaxy build up in numbers for a certain period of time and the time which is available will tend to be longer for the highly charged heavy nuclei.

The problem with this model is making it fit the detailed observations. Since iron nuclei have radii of curvature which are twenty-six times less than protons of the same energy and the radii themselves are proportional to the energy for a given nucleus, then we would expect an 'iron knee' to occur at an energy twenty-six times higher than a 'proton knee'. In between, there would be knees for all the particles with charges between 1 and 26. If the sources of all particles are the same, then we should be able to predict with confidence both the detailed shape of the spectrum and the progressive change in the composition. The model does not seem to work correctly at this level of detail. The observed knee has the wrong structure. It seems to be too abrupt, not spreading by twenty-six times in energy.

So, just when we were doing well, we have a setback. We still do not have a convincing answer to the origin of the knee. We feel that we really must be on the right track, the physical ideas seem straightforward and compelling—but Nature is really rather more complex than we might wish. Of course, we also have the problem of finding where the cosmic rays come from when their energies are well above those of the knee. The popular view is that their origin is outside the galaxy in the Cosmos but, at present, we can't confidently identify such a place.

## Cosmic ray arrival directions

We've seen that cosmic rays, being charged particles, are unable to travel in straight lines through the magnetic fields which fill our galaxy. Radio astronomers have also detected magnetic fields in intergalactic space. Cosmic rays will be deflected by these fields and the directions of their origins will be lost except maybe for rays of the highest energy (and lowest deflections). There may still be some preferred directions but these will be spread over broad regions of the sky due to the effects of the magnetic fields. The directional effects we find are amazingly small, the deviations from uniform brightness over the cosmic ray sky are almost always less than 1 per cent and often less than one-tenth of that value. This very smallness is so unexpected that it makes its study very important—and difficult. If Nature gave us sources of cosmic rays within the Milky Way, how could it spread their directions so efficiently that they arrive more uniformly at Earth than the light reflected from a sheet of good quality white paper?

One explanation suggests that cosmic rays fill our galaxy, or perhaps even the Universe. As a first guess, we might imagine that their average properties (number per cubic metre etc.) are quite well fixed over periods of time. For instance, we would not expect the galaxy to periodically fill and then empty of cosmic rays. On the other hand, the galaxy might be like a closed room full of air. Such a room contains many molecules all moving at high speed in random directions but, on an average, having no particular motion or direction. In our closed room full of air, we would not identify any particular arrival direction for the air molecules. The molecules would arrive uniformly in all directions, that is, isotropically. Cosmic ray physicists often search for deviations from isotropy; they search for a so-called anisotropy.

In this closed room, we would not feel excess pressure from any particular direction. Suppose now that we move our hand through the air. We will recognise that the air exists because we will feel some pressure on the forward side of our hands. If we had pressure sensors on our hands, we would measure an overall variation in pressure from the forward side to the reverse. This would go through one cycle from high to low and back to high if we were to move the sensors once around our hands. We would certainly detect a preferred direction for the arrival of the gas molecules onto our moving hands. This would be the direction of the highest pressure but it would be only a very general direction, defined by the single cycle of variation around our hands. Why is this preferred direction the one with the highest pressure? This is just because the gas pressure measures the number of molecules hitting our hands in a given time and this increases in the forward direction as we move our hands in the gas. The reverse surfaces of the hands are always moving away from the molecules; they are always trying to catch up and not all of them manage to do so. Thus, the pressure there is less. In cosmic ray terms, we would say that there is an anisotropy in the direction of the highest pressure.

How does this relate to cosmic ray anisotropies? Well, there is nothing static in the Universe. Our Earth travels at high speed in its orbit around the Sun, the Sun travels at an even higher speed around our galaxy and the galaxy itself is also moving past other galaxies in the Universe. We might reasonably expect some anisotropy of cosmic rays associated with these motions. But how much might this be?

This particular form of anisotropy is called the Compton–Getting effect. The expected value of the resulting anisotropy will depend on the speed of the moving body past the general mass of the cosmic rays themselves. Its actual technical value also depends somewhat on the form of the cosmic ray energy spectrum. Our solar system moves at about 200 kilometres per second through the galaxy and the resulting expected difference between the strongest and weakest directions for the cosmic ray intensity is about 0.1 per cent. Disappointingly, although delicate experiments have now reached these levels of sensitivity, no Compton–Getting effect has been convincingly observed even though it must exist at some level.

Importantly, there is another effect to look for. No doubt many cosmic rays originate within our own galaxy and progressively wander (diffuse) out into intergalactic space. This process is rather like having a draught in our closed room of gas molecules. On an average, these cosmic rays are expected to be moving past us in their outward journey which probably follows the spiral arms of our galaxy, guided by the magnetic fields.

We have been able to measure this effect at low energies but at a departure from isotropy well below 1 per cent. Such a measurement requires huge numbers of cosmic rays to be recorded since the process involved in the experiment is statistical and the statistical variations have to be averaged out. The effect is very small and a large data set is required in the same way that an opinion poll would have to be huge to get a result which is valid at the 0.1 per cent level. Such event numbers are only possible at the lower cosmic ray energies and we are left to speculate about what happens at higher energies. For instance, we expect some fundamental changes to occur near the spectrum knee and it has been hoped that they might result in measurable anisotropies. So far, our experiments have not been sensitive enough to measure this effect. At the very highest energies, we surely should have an effect but, apart from an intriguing result we will discuss later, present observations have proved unsuccessful, so much so that we have every reason to begin to question our understanding of the high energy Universe.

Strangely, there is another astrophysical phenomenon, the gamma-ray burst, which presents us with an almost identical problem in terms of source directions to that of the cosmic rays. It is intriguing to speculate that the sources might be related. We discuss this in the next chapter.

# CHAPTER 7

# COSMIC GAMMA-RAYS

Until the end of the 1970s, cosmic ray research was almost solely the study of energetic, charged cosmic ray particles. In the 1960s, the study of the skies by cosmic ray scientists using X-rays rather than light or radio waves had become possible with the development of instrumentation for high altitude balloons and rockets. This new astronomy was an exciting offshoot of cosmic ray studies but soon went its own way. Cosmic ray research itself moved into the doldrums. There were interesting things to study but they didn't seem to be moving us towards breakthroughs. Some scientists in the field deliberately searched for the strange and bizarre such as particles travelling faster than light or exploding black holes. We need to keep open minds in physics and part of the role of physics research is to investigate even the most unlikely possibilities where we can. However, these investigations were, at best, exciting sidelines and, on the whole, cosmic ray studies were starting to look a bit worn out. All that changed in 1982 because of an extraordinary discovery in Germany. Since then we haven't looked back. Cosmic ray research is now one of the great frontiers of modern astronomy.

The big problem with finding where cosmic rays come from is that they do not travel in straight lines. The Universe seems to be full of magnetic fields which change the direction of travel of the cosmic rays over the huge distances from their sources to us. Some particles—those without an electrical charge—don't have this problem.

Neutrinos are one such particle type but they are very difficult to study since they interact so little with material that it is almost impossible to detect them. Billions of neutrinos would pass through a detector for every one that could be recorded. Still, astronomy using neutrinos is actually possible. It is in its infancy but holds great promise.

Another uncharged particle is the neutron. This is a particle much like the protons found in cosmic rays except that it has zero total charge and so is not deviated from its original path by magnetic fields. It would be an ideal particle to study for cosmic ray astronomy except that a neutron on its own will only live for about 15 minutes before falling apart to give a proton and an electron in a process called beta decay. In a time of 15 minutes, a particle could travel, at best, only a very short distance in astronomical terms—only about twice the distance from the Earth to the Sun. One would think that such a particle would not even reach us from outside the solar system. Surprisingly, this is not quite true. Very high energy neutrons in fact could travel large distances. This is because of relativistic time dilation (see Appendix 1) which is the effect that allows muons to travel through our atmosphere without decaying. A neutron with an energy of $10^{15}$eV (a million times its rest mass) would have its lifetime extended up to 30 years by the same factor of one million, and so it could travel almost 30 light-years. Even more energetic neutrons could reach us from anywhere within our galaxy. The third possible particle which travels in a straight line is, of course, a gamma-ray. This is simply an energetic form of light and travels in a straight line just as light does.

The 1982 announcement that rekindled the study of cosmic rays was made by researchers at the University of Kiel in Germany. Using a cosmic ray air shower array, they detected a cosmic ray signal at an energy of $10^{15}$eV from the well-known X-ray source Cygnus X-3. This announcement was shattering for those elsewhere in the field. No-one had expected to see such a signal, partly because it was assumed that the vast majority of cosmic rays were charged particles which would not travel in straight lines from their source. The possibility that very energetic gamma-rays were producing the showers was just as surprising because these energies were a million times higher than for any other known gamma-ray.

Cygnus X–3 is a strong source of X-rays in the constellation of Cygnus. It is believed to be a binary system of two stars, one a conventional star and the other a neutron star. They each orbit around the system's centre of mass once every 4.8 hours. The X-rays probably come from material which flows in a stream towards the dense neutron star, sucked in by its intense gravity from the tenuous outer regions of the conventional star. This fast-moving 'accreting' material heats up by friction as it approaches the neutron star and the hot material gives off X-rays. The Kiel researchers were using a fairly conventional cosmic ray air shower detection array at energies of about $10^{15}$eV and used fast timing (discussed in Chapter 5) to find the direction of each shower. They picked out all the showers from the general direction of Cygnus X–3 which they had recorded over a number of years and worked out when, within a cycle of the 4.8-hour Cygnus X–3 orbit, each had arrived. They then looked to see if any part of the cycle had an unusual excess of events. It did. This meant that some sort of particle had travelled from Cygnus X–3 to the Earth and retained its original direction plus its relative time. This could only be a particle which was not deviated by magnetic fields and had thus travelled in a straight line. It was assumed to be a gamma-ray. This was the birth of Ultra-High Energy (UHE) gamma-ray astronomy.

The Kiel observation immediately opened a new vista since other X-ray binary systems were known and could be searched for in data sets accumulated over the years by other research groups. Also, the new observations gave a way of studying directly the properties and interactions of gamma-rays at these energies. Gamma-rays entering our atmosphere would produce particle cascades in much the same way as cosmic rays produce extensive air showers. The expected big difference would be that the gamma-ray showers would have few muons since the muons in showers come mainly from the decay of pions which themselves result from the nuclear interactions of primary cosmic ray nuclei. Disturbingly, the Kiel data showed no clear difference between the muon content of the gamma-ray showers and the other showers they recorded. It was argued that this may not be a huge problem since there were mechanisms which might apply at these extreme energies to produce muons in gamma-ray showers. Still, the situation felt rather unsatisfactory. On the other hand, the Kiel observation was quickly confirmed by other detectors and signals were found from the directions of other X-ray binaries. All seemed well.

Over the next few years the field blossomed. More observations were made of several sources, although Cygnus X–3 remained the clearest beacon. Unfortunately, as time passed, the worries of the researchers increased. Physicists are by nature a sceptical group of people who quickly home in on loose ends. When the initial euphoria wore off a little they found disturbing inconsistencies. The muon issue was a worry, some of the claimed observations were inconsistent with others, the signal levels weren't always consistent and neither were the times in the orbital cycles when the signals were seen. On their own, each of these was not a great problem. The observations were difficult and the arrays had not been designed specifically for astronomy so a few inconsistencies were to be expected. However, there was a developing worry as usually when data are added together the overall collection of data becomes more significant and the results clearer. This wasn't happening. The big worry then became Cygnus X–3 itself. Its signal was fading.

Stars like our Sun burn their nuclear fuel in an orderly way for billions of years and have an in-built mechanism which ensures that their brightness stays pretty constant over those huge times. More bizarre objects in astrophysics generally lack that sort of stability and we commonly see variability in radio and X-ray astronomy. We might have expected Cygnus X–3 to have varied its signal quite strongly with time. On the other hand, it is not a happy situation to have your most convincing source progressively disappear! It was even worse since the disappearance happened at the same time as we learned to be more and more critical about our experimental technique and we built better detectors for studying the gamma-ray showers. There was no doubt that some apparently positive observations had been statistical accidents in the data, but could they all have been? That really was a frightening thought which critics of the field were not reticent in suggesting.

The present situation is that no-one can be absolutely sure whether or not the field of UHE gamma-ray astronomy really exists. Any field which depends on careful statistical analyses of data is susceptible to any deficiencies in allowing for the various random and systematic uncertainties in the data sets. There remain a number of observations which seem convincing and consistent with each other but there is a broad spectrum of quality in the observations and some (perhaps many) will finally prove to be incorrect. The finest instrument currently available for this work, CASA-MIA (the Chicago Air Shower Array and the Michigan Anti-Coincidence), is a very sophisticated combination of a large air shower array and muon detector built in Utah by a group from the Universities of Chicago and Michigan. This team is led by Nobel laureate, Jim Cronin. It has so far been unable to detect any clear gamma-ray signal from the northern sky.

One observation which we find encouraging but not yet completely convincing is that of our nearest active galaxy, known to radio astronomers as Centaurus A and to the optical astronomers as NGC 5128. It is a southern object and two groups claim to have observed UHE gamma-rays coming from it. They are JANZOS, a Japanese/New Zealand/Australian collaboration with an array (recently closed down) high on a mountainside in New Zealand and the cosmic ray group at Adelaide with an array at Buckland Park, just north of the city. 'Cen A' is just the sort of place we would expect to produce cosmic rays and secondary gamma-rays (through cosmic ray interactions in the outer galaxy). Also, it seems to have gamma-rays with energies predominantly below $10^{14}$eV and this is encouraging since the microwave background from the Big Bang should interact with and destroy gamma-rays a little over this energy if they travel intergalactic distances.

We take further heart in the fact that, in recent years, a form of gamma-ray astronomy at somewhat lower energies has become firmly established. This is known as Very High Energy (VHE) gamma-ray astronomy and operates at energies of about $10^{12}$eV, about one hundred times lower than the energies observed by the Kiel array. That is, intermediate between the highest energies studied using satellite-borne instruments and those we have just discussed.

VHE gamma-ray astronomy can be traced back to pioneers in the 1950s. Fundamental measurements at Harwell in England at that time showed that Cerenkov light was emitted by extensive air showers and that the light could be readily observed on clear moonless nights using sensitive photomultiplier tubes. This light arrives at the ground in very short bursts at close to the same time as the shower particles. The bursts last a few billionths of a second and for that brief time can be much brighter than the brightness of the starry sky. A simple light-detecting photomultiplier just looking at the night sky is sensitive to cosmic ray showers with energies of about $10^{15}$eV. Pulses of light from these showers can be recorded with ease. Lower energies can be studied if a mirror is used to collect the light onto the photomultiplier. In the 1950s and even later, searchlight mirrors were readily available and were popular for this purpose. We find it hard to believe now, but in Sydney, one of the world's great cities, this work was done close to the city centre. Clearly, light pollution is a child of the most recent decades.

It turns out that there are great advantages for gamma-ray astronomers to observe at as low an energy as possible. For ground-based telescopes, this seems to be in the range of energy just below $10^{12}$eV. To observe at such relatively low energies, one has to collect as much light as possible by using large mirrors and one has to use as many tricks of the trade as one can to enhance the gamma-ray signal against a large background of cosmic ray signals. A number of solar power stations have huge arrays of mirrors and some of these have been (and are being) used for this work which, of course, can be done when the power station is unable to carry out its designated purpose. Some years ago, the University of Adelaide used the prototype solar power system developed by the Australian National University for White Cliffs, a small, opal-mining town in outback New South Wales. These systems have a wonderful collecting power for Cerenkov light but their use has turned out to be limited since we now know that success in this field requires subtlety as well as the brute force of building large mirrors. Fantastic photomultiplier cameras are now used to capture the images of the showers with exposure times of a few billionths of a second, and the rather crude mirrors of the solar arrays are not often up to the task of producing sufficiently high quality images.

Many research groups are now working in this field of VHE gamma-ray astronomy using the Cerenkov technique, led by the US–Irish collaboration with a large 10-metre-diameter light-collecting mirror at Mt Hopkins in the south of Arizona. This group, which can trace its history back to the Harwell pioneers of the 1950s, has successfully developed a powerful technique for rejecting images of showers initiated by cosmic rays in favour of those from gamma-ray primary particles. Cosmic ray showers have a ragged image due to the muons. On the other hand, gamma-ray showers have a much cleaner tear-shaped image which points to the direction of the source to be studied. By making carefully chosen measurements of the images, cuts can be made in the data set which remove most of the ragged cosmic ray showers and improve enormously the ratio of gamma-ray to background events. With a bit of experimentation and the genius of Professor Michael Hillas of the University of Leeds, this group found an absolutely clear signal from the Crab Nebula, the best-known supernova remnant. They then went on to make an even more crucial observation. They obtained a clear signal from the direction of Markarian 421, the nearest of the class of energetic galaxy objects which includes quasars and active galactic nuclei. Both signals are so clear that there is no doubt that they exist. With Markarian 421 it is now possible to follow the large variations in its gamma-ray brightness on a daily basis. Other research groups (such as the Australian–Japanese collaboration, CANGAROO [an imaginative acronym: Collaboration between Australia and Nippon for a Gamma Ray Observatory in the Outback], based at the remote Australian township of Woomera), mainly using similar photomultiplier cameras, have now painstakingly added to the catalogue of VHE gamma-ray sources. A genuinely new form of astronomy has formed a solid foundation.

Gamma-rays at these energies and above are almost certainly produced in their astrophysical sources by processes similar to those operating in a cosmic ray shower. Recall that the cosmic ray enters our atmosphere and interacts with an atmospheric nucleus. This results in the emission of pions and the neutral pions decay to make gamma-rays. If there is a cosmic ray particle which passes through an astrophysical region containing gas, dust or even other radiation, it will interact in just the same way and the result will be that gamma-rays will come from that object. This is why we believe UHE gamma-rays come from neutron star binaries where the atmosphere of the conventional star is the target material. It is also how gamma-rays might be produced in the even more violent environment of the black hole which is central to a powerful active galactic nucleus. The Markarian 421 observation thus gives us incontrovertible proof that the sort of astronomical objects which we believe are the most energetic in nature have strong populations of high energy cosmic rays.

### Gamma-ray bursts

Cosmic rays come to us from all directions with no apparently preferred direction. There is another phenomenon in astrophysics which is also highly energetic, comes from all directions and is stubbornly mysterious. This is the series of strong bursts of gamma-rays commonly observed on spacecraft. These gamma-rays have energies typically about 1 MeV, much lower than those we have just discussed, but it is often speculated that they might still reasonably be related to cosmic rays in some way not yet understood.

Our search to understand the Universe and its contents has led us to discover many strange phenomena. For most of these, we have been ingenious enough to reach some understanding with the help of careful observation. Our apparent understanding of some may well prove to be transitory as our knowledge increases, but at least we are able to suggest some explanations which seem plausible to many people. In the case of gamma-ray bursts, however, we have a phenomenon which has been known for a quarter of a century and for which we have no significantly better explanation than our first tentative theories. In fact, for some time we had more theories than the number of bursts they were supposed to explain! Perhaps we should remember that science only *assumes* that our Universe is describable in a way which appears rational to us.

Gamma-ray bursts belong in a discussion of cosmic rays since we know that gamma-rays in the Cosmos are often the result of cosmic ray interactions at their sources. The explanation of the bursts could be a significant key to the mysteries of cosmic radiation. Also, these bursts have some of the same perplexing characteristics as cosmic rays. They come uniformly from all directions and we see no relationship at all between them and any other known astrophysical processes.

**Burst observations**

A characteristic of a nuclear bomb explosion is that a short but intense burst of gamma-rays is produced. An effective way of checking compliance with the 1963 Partial Test Ban Treaty was for the US military to launch the Vela series of satellites carrying gamma-ray detectors with detection systems arranged to respond to short bursts of gamma-rays. Fortunately, the satellites were sensitive to gamma-rays coming from space as well as from the Earth's surface. Unfortunately, when a burst arrived they were unable to say where it came from! Intense bursts apparently with the characteristics of nuclear explosions were discovered but were not made public for several years. No explosions were detected at the same times by more conventional methods.

At that time, in the late 1960s, Stirling Colgate in New Mexico and his colleagues were carrying out calculations on the detailed processes to be expected for various types of supernova explosions. One of their predictions was that a supernova would produce an intense burst of gamma-rays and this seemed to be a plausible explanation of the Vela phenomenon. However, data from the satellites were searched for any sign of bursts which arrived at the times of reported supernovae and no correlations were found.

In the absence of a ready explanation, it became necessary to examine the burst phenomenon from first principles in a systematic way. A more simple question to ask is not whether the bursts were specifically from super-novae but whether they were detected by more than one spacecraft at the same time, 'in coincidence'. With encour-agement from Colgate and Edward Teller (perhaps better known for his interests in nuclear bombs as sources of gamma-rays), a search was made using data from four of the Vela satellites in orbit from 1969 to 1972. The possible rate of bursts detected by any one of the spacecraft was rather low yet, over that period, sixteen bursts were observed by pairs of spacecraft and two were recorded by all four. The chance that this was an accident of statistics was unrealistically low. The conclusion had to be reached that the Cosmos was producing bursts of gamma-rays and that their unknown sources did not seem to be supernovae.

The bursts that are currently being observed and studied are mainly being collected by an experiment (the Burst and Transient Source Experiment or BATSE) spe-cifically designed for this purpose and part of NASA's space-borne Compton Gamma Ray Observatory which was deployed by the Space Shuttle in 1991. This experiment is very sensitive and detects bursts at a very much higher rate than did the Vela satellites. The bursts are made up almost entirely of gamma-rays and energetic X-rays. That is, most of the energy is carried by particles with energies of 1 MeV or so and very few of the particles have energies below 0.05 MeV. No optical counterpart of a burst has yet been found despite considerable effort to search the records of observatories.

Bursts are detected at an average rate of roughly one per day. As we said, the bursts are short-lived, usually lasting only a few seconds although they have been observed to be as short as thirty-thousandths of a second or as long as 100 seconds. It is intriguing that there seem to be too few bursts lasting about 2 seconds, and it may be that there are two different types of bursts—short ones of a second or so duration (an average of 0.3 seconds) and longer ones with durations longer than 3 seconds (an average of 20 seconds). We have to be a bit careful here because it could be that if our instruments were sensitive enough there might be signals lasting much longer. This is suggested by a very intense burst on 17 February 1994 which at first appeared to last 180 seconds but was later found to have shown effects up to 10 hours later. Astronomers often try to see how fine the time structure is in the phenomena which they observe since, as we have seen, this gives a clue to how big the source might be. Time structure shorter than one-thousandth of a second has been seen in a burst which suggests a tiny source with a size below 300 kilometres.

We will soon see that an accurate way of finding the direction of a burst is to measure the times at which it passes a number of satellite detectors. BATSE is rather different to previous burst experiments in this sense, as it is capable of determining with reasonable accuracy the direction of each burst using just its own detectors. Potentially, this is a very powerful facility because if Earth-based observatories could be informed of the burst direction as it occurred, it would be possible to immediately search that region of the sky with optical, radio or other telescopes to identify the source. This process is almost possible with

BATSE but it still has serious limitations due to its rather large uncertainty in direction and the time still necessary to process the data. The next generation of burst experiments may be needed before this aim is fully achieved. A small satellite experiment called HETE (the High Energy Transient Experiment) being developed in the USA may soon make such a multi-telescope experiment practicable for the first time.

From very early times, the observed bursts were often found to begin very suddenly, and with the timekeeping available on the spacecraft the time at which a burst arrived at a particular spacecraft could often be specified to within about one-twentieth of a second. This timekeeping can allow one to find the direction from which the burst arrived in much the same way that we reconstruct the directions of cosmic ray showers in the atmosphere using an array of detectors on the ground. The orbits of the Vela satellites had a radius of a little over 100 000 kilometres. If they happened to be as far apart in their orbits as possible, two satellites could be spaced by twice that distance when a burst occurred. Since gamma-rays are a form of high energy light, they travel at the speed of light—300 000 kilometres per second. This means that it could take as much as two-thirds of a second for a burst of gamma-rays to travel to a second satellite after hitting the first. Of course, this time depends on the exact arrival direction of the burst relative to the satellites and the spacing of the satellites. Since the positions of the satellites are always known very accurately, the time difference can be used to calculate the direction of the source of the burst. At least three satellites are needed for this to give really useful directions. The more the better. For instance, if the burst was to come exactly perpendicular to the line joining the satellites, the time difference would be zero. Any other orientation would give an intermediate time difference up to a maximum of two-thirds of a second which would mean that the burst had come from a direction along the line joining the satellites.

This technique was clearly rather crude when only three

Vela satellites were used, but, as time went by, any gamma-ray detectors in other spacecraft were pressed into the search and some of these were at very large distances from the Earth, thus making some burst directions very well defined. For instance, soon after the discovery of bursts was announced, a burst was detected on the Vela craft and on the Apollo 16 command module. The module's gamma-ray detector was actually meant to be used for determining the composition of the lunar surface but it was also capable of detecting gamma-rays from anywhere else. In this early case, the direction of the source was located with an uncertainty of 15 degrees. Such a directional uncertainty is too big to compare with optical astronomical photographs but the large angular distance of 50 degrees from the direction of the plane of the Milky Way was to be a portent for future worries. This suggested that the bursts were not from our galaxy. If they were, then they must have been from very close, from within the thickness of our spiral arm.

It soon became clear that there were, indeed, no directions in the sky which were major sources for the bursts. Even with the crude direction finding from the early experiments, there was no clustering of bursts around the direction of the Sun or in the general direction of the Milky Way or our closest neighbouring galaxies. This remains true twenty years later and is a major mystery in astrophysics.

Astronomers have learned to be patient under such circumstances; clues to the distances of cosmic sources have usually come along with the passage of a little time. It is less than a century since we found a realistic size for our galaxy and realistic distances to our galactic neighbours. Sometimes there is a lucky break, as occurred with pulsars which have radio waves which progressively spread in time and rotate as they pass through our galactic magnetic fields and the ionised gas within them. In that case, measurements of the spread and the rotation of the radio waves give a measure of the distance of the pulsar from the Earth (and a bonus measurement of the magnetic field). Radio astronomers knew about pulsar distances almost from the first discovery of the phenomenon (curiously at about the same time as the discovery of gamma-ray bursts). Luck of that sort has not been with us for gamma-ray bursts and we are forced to use any clues we can find in deducing their distances and possible sources.

There is a particular technique which tends to be brought in as almost a last resort under these circumstances. It involves looking at how the number of bursts varies as a function of brightness.

The apparent brightness of a star or, in this case, the intensity of a burst of gamma-rays decreases as one moves further from the source. This is our common experience. The exact mathematical way in which the decrease occurs is known as the inverse square law. It follows from the fact that as a light beam moves away from a star, the light covers a greater and greater area. For a star which shines light in all directions, doubling the distance to the star will reduce the brightness four times, since the area of the sphere now illuminated by the light has increased four times (the area of a sphere increases with the square of its radius). Mathematically then, the apparent brightness (the amount of light per square metre) decreases with the square of the distance.

Suppose now that we are astronomers making measurements of the brightness of stars and suppose for the moment that the true brightness (the total number of photons emitted per second) of all stars is the same. This is not true, of course, but the argument can be shown to still work. When we look at the real sky, we still see much the same sort of thing as we do in this simple model. Some stars are bright and others faint but this is mostly because some stars are close and others distant. The question is, how many stars are there in our simplified model with greater than a certain brightness? Again, we will ask how things change if we double the distance to a star. As we saw, the brightness is reduced to one quarter of its original value but the volume out to the distance of that star has increased by eight times (the volume of a sphere depends on the cube of the distance). There are now eight times as many stars closer to us than our star and all appearing to be brighter because they are closer. We can see that there is a relationship between the brightness which we choose and the number of brighter stars.

We plot a logarithmic graph of brightness versus the number of brighter bursts. This method should be really quite powerful since the steepness of the graph is given by the value 3/2 (the 3 part of this comes from the cubic relationship we just saw for the volume and the 2 part comes from the inverse square law) which, in turn, depends on how the stars (or gamma-ray sources) are spread in space. Notice that the cube part of the argument was from the relationship between the volume and the size of a sphere. If our stars had not been in a spherical volume, that relationship would have been different. If we had used a disk (like a flat spiral galaxy), this would have been squared and not cubed. If the stars had been in a line (like along a spiral arm of the galaxy), volume and distance would have been proportional. Our experiment to find information on the source of gamma-ray bursts works like the argument above for stars. We find how many bursts there were above various brightness levels and see whether this fits the curve for a sphere (like the Universe as a whole or just a tiny local spherical region), a disk (like sources spread through the plane of our galaxy) or a line (like a spiral arm).

When this measurement is done, the answer is intriguing. The graph does indeed have the slope of 3/2 but only for the rarer, strong bursts. At the end of the graph which represents the weak bursts, there is a deficit. It is always difficult to make measurements like this (in fact, for technical reasons which remove some bias in the way bursts are selected, even this experiment has to be a slight modification of the one we described). However, the BATSE curve fits well with another from the Pioneer Venus Orbiter spacecraft which has collected data for much longer (10 years) and so we can have reasonable confidence in the result. This would seem to mean that the brighter (nearby) burst sources are distributed uniformly around us but that at some distance there is an end to the volume which results in us seeing too few distant sources. The problem is that we don't know how big the source volume is. How far away is its edge? It has been thought at one time or another that this edge could be the edge of our local solar region (the heliosphere or the cometary cloud around the Sun), the edge of our galactic halo (ten thousand times bigger) or even the edge of the Universe (a million times bigger still). So we have learned some things but we are still without some vital clues.

It is obvious that these gamma-ray bursts are an important mystery in astrophysics. When they were first observed, it was thought that they were probably associated with effects near neutron stars in our galaxy. This idea was popular for fifteen years or so but now appears unlikely as it has difficulty explaining the intensity and directional distributions. We still have no common agreement about their possible origins. However, our experience is that the processes which produce gamma-rays are closely related to those which produce cosmic rays. There is reason to hope that there may yet emerge some common solution to the origins of both the high energy cosmic rays and the gamma-ray bursts.

# CHAPTER 8

# THE HIGHEST ENERGY COSMIC RAYS AND THE FLY'S EYE

Detecting the highest energy particles in the Universe is a technical challenge. Increasingly, astrophysics experiments are taking to the skies as payloads on high altitude balloons or satellites. If your interest is X-ray or gamma-ray emissions from stars and galaxies, this makes a lot of sense. A satellite or balloon gets you above the Earth's atmosphere, a strong absorber of this sort of radiation. On the other hand, if you wanted to catch ultra-high energy cosmic rays this way, you would need to be ultra-patient. At energies above $10^{19}$ eV, an average of one cosmic ray arrives in an area of 1 square kilometre every year. Put another way, a typical satellite detector, with an area of 1 square metre, would see one of these cosmic rays every million years!

You would think that scientists studying these cosmic rays would despair at the rarity of their quarry. But they take comfort in the fact that they have one up on some of their colleagues, such as those building huge detectors to observe gravitational waves from collapsing stars, or detectors to observe high energy neutrinos from active galaxies. Cosmic rays of enormous energies are known to exist, and are detectable! The same can't be said for gravitational waves or energetic neutrinos. Over a period of more than thirty years, a handful of cosmic ray particles with energies greater than $10^{20}$ eV have been observed. They haven't arrived from the directions some may have expected. In fact, as time has progressed we have realised the difficulty of tracking down the origin of these particles. Despite much progress in our understanding, which we'll describe in this chapter, there is still some way to go. As we'll see, the next stage in the quest will require an international effort comparable with the largest scientific projects ever proposed. Only then will we know how a proton, with a mass of one billionth of a billionth of a billionth of a kilogram can be given the same energy as a brick falling from the roof of a one-storey house.

Rather than cursing the atmosphere as an absorber of radiation, we've seen how cosmic ray physicists use it to their advantage. Except for those interested in the lowest energy cosmic rays, where the particles are copious enough to be seen in small balloon and satellite detectors, scientists use the atmosphere to make the rare particles more visible. The extensive air showers produced by cosmic rays in the atmosphere convert the energy of the primary particle into a large number of secondary energetic particles. The particles cause the air to glow in several ways (notably Cerenkov and fluorescent light) which allow the cosmic rays to be detected remotely. The shower spreads out in the classic saucer-shaped front that allows the cosmic ray to be detected with an array of particle detectors at ground level. Sampling the air shower at a number of points on the ground can be sufficient to determine what we need to know about the original cosmic ray—its arrival direction, its energy and its mass.

## The first giant array

Since Pierre Auger's discovery of extensive air showers, scientists have been building larger and larger arrays of detectors in some very exotic and inhospitable locations around the world. But it wasn't until the early 1960s that a large enough array was built to tackle the origin of the highest energy particles, those with energies above $10^{17}$ eV. Bruno Rossi's prolific group from the Massachusetts Institute of Technology built a dedicated observatory, after making important contributions to the techniques of measuring air showers with scintillation detectors. In a remote area of New Mexico at Volcano Ranch, the new array was built and operated by a team led by John Linsley. This project, the first of the giant arrays, consisted of nineteen detectors, each with an area of 3.3 square metres, spread over a ground area of 8 square kilometres. Volcano Ranch was to operate for a total of three years in this early form, collecting 1000 showers with energies above $10^{18}$ eV and making fundamental contributions to our knowledge base.

For example, it discovered that even the ultra-high energy cosmic rays did not arrive from any preferred direction. In other words, as far as Linsley could tell, the arrival directions were isotropic. This was a surprise for many people at the time, though it was understood and expected at lower cosmic ray energies. It all has to do with how moving charged particles interact with magnetic fields.

As we've seen, it's the spiralling of charged cosmic rays around the tangled magnetic fields in our Milky Way galaxy that makes it impossible for us to trace back the arrival direction of a typical cosmic ray to its astronomical source. However, we expect things to be very different for the highest energy particles. The amount of bending experienced by a moving charged particle in a magnetic field is proportional to the strength of the magnetic field and the charge of the particle. In addition, the bending reduces as the energy of the particle increases. So as we look at higher and higher energy particles, they become rarer, but they also move in straighter and straighter lines. Indeed, the expectation for the Volcano Ranch data was that it might well show a clustering of cosmic rays arriving from the direction of the band of the Milky Way. That this was not observed could be interpreted as an indication that cosmic rays did not originate in our galaxy. However, the relatively small number of showers, especially at the highest energies where the particle trajectories would be the straightest, meant that any conclusion couldn't be very firm.

Important as the isotropy result was, Linsley made a much more exciting discovery with his array. A particular air shower landed one day on the array. This shower was unusual in that a large number of shower particles were

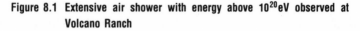

**Figure 8.1 Extensive air shower with energy above $10^{20}$ eV observed at Volcano Ranch**

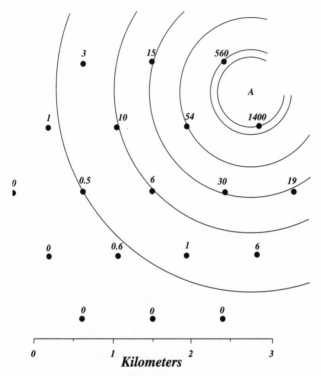

A map of the first extensive air shower with an energy above $10^{20}$ eV observed by the Volcano Ranch array. Each dot represents the position of a detector on the ground and the numbers refer to the observed particle numbers per square metre. Position A represents the location of the core of the air shower.

_____

detected in widely spread detectors. While a typical shower would deposit particles in four or five detectors, this shower was seen by fifteen detectors, with larger than usual particle numbers. A careful analysis of this single event showed that the shower had been initiated by a cosmic ray with an

124

energy in excess of $10^{20}$ eV, the most energetic particle observed at that time, and 100 000 times more energetic than the particles detected with Auger's pioneering air shower experiments. Details of Linsley's event were immediately published in the prestigious journal *Physical Review Letters* in 1963 and caused considerable interest. The remarkable nature of this cosmic ray was emphasised three years later, in 1966. It was realised that cosmic rays of this energy would strongly interact with the cooled-down radiation left over from the fireball that was the Big Bang.

## Cosmic rays and the microwave background

It was only a year after the discovery of the cosmic microwave background in 1965 that Kenneth Greisen in the United States and Georgi Zatsepin in the Soviet Union put forward a new theory that has had a major influence on Linsley and his successors. Greisen and Zatsepin realised that cosmic ray protons with energies above approximately $6 \times 10^{19}$ eV will suffer catastrophic collisions with the microwave background, and through each of these collisions the protons would lose significant fractions of their energy. The two scientists were making use of extensive experimental data on collisions between protons and photons of electromagnetic radiation collected in controlled experiments on Earth.

How did they make the connection between experiments made with very modest proton energies and the enormous power of the highest energy cosmic rays? The basic physics involved is familiar. It's the Doppler effect, recognised by all of us when the sound of an approaching train whistle appears to increase in pitch. In a similar way, when a highly relativistic cosmic ray proton rushes towards a low energy, long-wavelength photon of the microwave background, the proton sees the wavelength of the photon decrease until, as far as the proton is concerned, the microwave photon has transformed into a gamma-ray! This effect is described as the relativistic blueshift of the photon. In this way, the collision is indistinguishable from a common laboratory experiment where physicists fire rather low energy protons towards gamma-rays. In the laboratory the result of the collision is a spray of fundamental particles including neutrinos and pions. Some of the combined energy of the proton and the gamma-ray converts to mass in the form of pions, which come in three varieties (positively charged, negatively charged and neutral) and have a mass about one hundred times larger than an electron. The collision in space between the higher energy cosmic ray proton and the feeble microwave photon causes the same spray of pions and neutrinos, and the original cosmic ray loses roughly 20 per cent of its original energy. Interestingly, in the collision it's possible that the proton could change into a neutron.

Greisen and Zatsepin realised that this effect would only rob energy from the highest energy cosmic rays. Only those protons with energies above the threshold energy of $6 \times 10^{19}$ eV would see microwave photons sufficiently blue-shifted to produce pions. Only these cosmic rays would lose energy in the collision. So while space is full of cosmic rays, and most sail happily through the sea of radiation called the microwave photons, the highest energy particles see the same radiation as something of a brick wall. On average, one of these cosmic rays will suffer a collision every 20 million years and lose 20 per cent of its energy. If the source of cosmic rays is close enough this won't pose a problem. But if sources are distant, say more than 150 million light-years away, this process will mean that we won't see any proton cosmic rays above the Greisen–Zatsepin threshold energy. Half a dozen or so collisions will have robbed them of their super energy. We now realise that other varieties of cosmic rays, those that are nuclei of elements, will also suffer catastrophic interactions at these high energies, this time through collisions with photons of starlight. The average time between collisions for these heavier cosmic rays is typically shorter than the 20 million years for their proton cousins.

Is a brick wall an appropriate metaphor for a collision that happens once in 20 million years? Perhaps not, but a distance of 20 million light-years is really very small in the context of the enormity of the Universe. Astronomers have viewed quasars at distances five hundred times larger, near the edge of the observable Universe, 10 billion light-years, away. The upshot of all this is very important. If we see cosmic rays with energies around $10^{20}$ eV, we are seeing particles accelerated in our neighbourhood of the Universe, say within 150 million light-years. Linsley's first event with this energy was therefore extremely important. It was not necessary to search the whole of the Universe for its origin. It must have been created in our 'backyard'.

The presence of the Greisen–Zatsepin effect and other similar processes makes the search for the origin of $10^{20}$ eV particles easier and more difficult at the same time. While we don't need to look as far for their source, we know of no super bright, quasar-type objects (an attractive guess for a source) in our region of the Universe. So while we don't have a theory for how quasars would accelerate cosmic rays to this enormous energy, it seems that they are removed from the competition anyway because of an effect that occurs in the emptiness of space. We must continue our search for a less obvious source, but one no less remarkable—perhaps a nearby active galaxy. After all, these particles must be accelerated somewhere!

### Giant arrays around the world

After Linsley's pioneering efforts, others were tempted to get into the game, especially with the carrot of Volcano Ranch's extraordinary event. Arrays were planned and built from the mid-1960s in the United Kingdom, the Soviet Union and Australia using a variety of techniques. The Haverah Park array, near Leeds in the north of England, was completed in 1968. It was 50 per cent larger than Volcano Ranch with a ground coverage of 12 square kilometres. The group was originally led by John Wilson (later by Alan Watson) and consisted of a collaboration of a number of UK universities including Leeds, Durham, Nottingham and London. Haverah Park used a new method to detect the air showers as they hit the ground. Instead of putting out detectors made out of plastic scintillator, they used large tanks of water in various positions in the array. A total of 550 square metres of detector area was spread around the rolling Yorkshire dales, involving a total of 600 tonnes of water. Water is a cheap and effective detector of particles travelling close to the speed of light, due to the Cerenkov effect.

We saw earlier that Cerenkov had found that when a charged particle moves quickly through a dielectric material (one made up from molecules that exhibit a slight charge imbalance across their breadth), it causes the molecules in the medium to emit light. Water, like air, is a dielectric material. More importantly, if the velocity of the particle is faster than the speed of light in the material, the light emitted by the molecules will add together in such a way as to form a strong shock-front of light. Since the energetic particles in an air shower are travelling close to the speed of light in a vacuum, and the speed of light in water is about 70 per cent of this value, a water tank will give off a strong, but extremely brief, flash of Cerenkov light when an air shower passes through it. The flash of bluish light can be as quick as 20 billionths of a second, and is picked up by sensitive photomultiplier tubes and converted to an electrical pulse.

Wilson, Watson and their colleagues exploited the Cerenkov effect, and built a very successful array of light-tight water tanks, built of thin galvanised steel that was no barrier to the high energy shower particles. Some say they were lucky that they used pure Yorkshire water and not a murkier variety that would have been corrosive and would have encouraged all sorts of undesirable life forms to grow in the warm, dark environment. Haverah Park ran for a total of twenty-three years, and at the emotional party held at the site in 1991 to turn the array off for the final time, a large group of past and present researchers on the experiment gathered to toast the occasion with a very refreshing drop. Not champagne, or a cleansing Yorkshire ale, but 23-year-old water from the central water tank! Fortunately the water was as fresh and sweet as ever—not tainted at all by the billions of shower electrons, muons and gamma-rays that had passed through it over the decades!

Another giant array was being built in the early 1960s. Halfway around the world from Haverah Park in the eastern provinces of the Soviet Union at Yakutsk, a group of physicists from Moscow State University was hard at work. The array combined elements of Volcano Ranch and Haverah Park, with an array of plastic scintillator detectors interspersed with an array of Cerenkov light detectors. By the mid-1970s the project covered an area of 20 square kilometres. The Cerenkov light detectors did not look for light in tanks of water—instead they looked for flashes of this bluish radiation in that other well-known and common dielectric material, air. The atmosphere was their detector, and an array of bare photomultiplier tubes pointed towards the sky on clear nights to detect the light from giant air showers.

You might think that it would be impossible to see the flash of light from an air shower amongst all of the background light from the stars and other atmospheric emissions including airglow. But these air showers produce an enormous amount of Cerenkov light because of the great number of relativistic particles they contain. The ability to look at showers in two ways, through the particles hitting the ground as well as through the light emitted high in the atmosphere, was a great boon. For one thing, it allowed scientists in Yakutsk and in Haverah Park to have an independent way of checking their methods for assigning energies to the original primary cosmic ray particle. As we'll see presently, the extra ability to see the shower develop in the atmosphere is very important.

Australian scientists were not idle during the active decade of the 1960s. They constructed a cosmic ray observatory that became the largest ever built, coming in at an enormous 70 square kilometres of ground coverage. Built by Brian McCusker and his Sydney University colleagues, the SUGAR array was situated in the Pilliga State

Forest near Narrabri in New South Wales. SUGAR was one of the first acronyms coined in high energy astrophysics and stood for the Sydney University Giant Airshower Recorder. Each of the 47 stations in the array consisted of two scintillator detectors buried under 2 metres of soil. Being buried, the detectors were sensitive to the penetrating component of air showers, the muons. This meant that the array as a whole was not as sensitive to the less interesting air showers produced by the more numerous lower energy cosmic rays.

The SUGAR array, because of its size, was forced to use some novel techniques for data collection. In fact, its innovations were to point the way to an array being planned right now. All previous arrays connected the detectors to a central data collection station with cables. Over the large distances and difficult terrain in the Pilliga State Forest, this wasn't practical. Instead, the Sydney scientists used state-of-the-art electronics to record the data on tape recorders in each of the 47 detectors, and used radio signals from a central site to keep the individual station clocks synchronised. SUGAR, like all the arrays to that time, measured the arrival direction of the shower by measuring the arrival time of the shower at stations spread across the array. For this reason, the synchronisation of the station clocks had to be better than 50 billionths of a second, a great technical challenge for the late 1960s but one that was achieved. SUGAR collected data for a total of eleven years from 1968 and remains the only array ever built in the southern hemisphere large enough to study cosmic rays with energies above $10^{17}$ eV. Among other legacies, we'll see that it left behind tantalising evidence of a source of cosmic rays in the Large Magellanic Cloud.

## A new type of detector—the Fly's Eye

The weather in the area around SUGAR, and in northern New South Wales in general, is noted for being fine and the atmospheric clarity is excellent. This has attracted several astronomical observatories to the vicinity including the Anglo-Australian Telescope and the predecessors of the Australia Telescope. In the late 1960s, it also attracted the thoughts of Kenneth Greisen of Cornell University, the same scientist who had predicted the interactions of high energy cosmic rays with the microwave background.

Greisen was considering moving his unusual new cosmic ray detector to a more suitable location. His team had been operating its 'Fly's Eye' detector in the cloudy and wet environs of Ithaca, New York, just a few kilometres from the university. The Fly's Eye was an optical experiment which took its name from its structural similarity to the multi-faceted optics of the insect's eye. Like the Yakutsk Cerenkov detectors, it searched for light emitted by air showers on moonless nights. Ithaca was almost the worst place in the world for this work and it made a lot of sense to relocate the experiment to the SUGAR site for some observations of showers with the two techniques. Unfortunately, a year after these thoughts, Greisen abandoned these plans when the ambitious Fly's Eye trials failed. The detector failed to see air showers. Yet these pioneering efforts were not wasted. The story of the Fly's Eye was not over and it has now evolved into the most versatile and sensitive cosmic ray detector ever built. To understand why Greisen initially failed and how these problems have been overcome, it is worth discussing the technique in some detail.

A. E. Chudakov of the Institute for Nuclear Research in Moscow was one of the pioneers in detecting Cerenkov light from air showers. His experiments in the 1950s led the way to the implementation of the Cerenkov array at Yakutsk and his ideas led to the birth of the Fly's Eye. One characteristic of Cerenkov light emission from air showers is that the blue light is emitted in a narrow cone around each relativistic particle in the shower. In air, the cone is only about a degree wide, meaning that when the Cerenkov light hits the ground, the light pool is very similar in size to the area hit by the shower particles. The great advantage of viewing the light, as well as the particles, is that the particles emit light during all stages of the shower's development. This makes it possible to extract some information about how the shower developed. As we've seen, this tells us more about the nature of the initial cosmic ray, particularly its mass, and it gives us a better estimate of the energy. The Cerenkov light, while short-lived, is extremely bright if you happen to have your detectors set up right under the shower. However, because of the very tight beaming of the light, it's not possible to view the shower from the side. In the 1950s, Chudakov realised that there was another source of light in air showers and that this light was emitted isotropically, that is, in all directions. If a detector could collect enough of this light, it would be possible to view air showers from the side, even from a large distance. The shower would not need to land on top of the detector for it to be seen. Chudakov saw that this opened the door to huge areas of ground being viewed by relatively small detectors.

A simple fluorescent light-fitting inspired Chudakov. When a current passes between the two terminals inside such a light, the electrons in the current collide with gas molecules, exciting electrons to high energy states. As those excited electrons return to their usual state, the excess energy is released as photons of light. This process is called fluorescence or scintillation. Chudakov pointed out that a cosmic ray air shower is just like an electric current—it's made up of moving charged particles, mainly electrons and positrons, moving through a gas, the air. Like many a good idea, it didn't appear to work when Chudakov tried to detect fluorescence from air showers with a small pilot experiment in 1956. The light emitted to the side of the showers seemed very weak indeed.

However, Chudakov's work inspired K. Suga from Tokyo University and Ken Greisen independently to tackle the topic in the early 1960s. One of Greisen's graduate students, Alan Bunner, who has since gone on to become a top NASA administrator, was given the task of finding out which components of the air were most likely to emit fluorescent light. He also set out to measure the strength of any emission. This effort called for some careful laboratory experiments where beams of fast electrons were passed through tanks of air of varying pressures and temperatures to simulate the passage of shower particles through the atmosphere. Given that an air shower typically starts 10 to 15 kilometres above the ground, Bunner needed to simulate the cold, rarefied lower stratosphere as well as the more familiar temperature and pressure near the ground. He identified that the most useful source of fluorescent light was from several excited states of molecular nitrogen, $N_2$, the most abundant gas in the atmosphere. The light is emitted just beyond the blue end of the visible spectrum in the near-ultraviolet region. Unfortunately, Bunner found that the light was only emitted very weakly, explaining Chudakov's early failure. Bunner saw that only four or five photons of ultraviolet light were emitted from every fast electron moving through a metre of air. Surprisingly, the amount of fluorescence did not depend too much on the pressure of the air or its temperature. High pressure air does not produce more light as might have been expected since collisions between neighbouring gas molecules are more likely in a denser gas and these collisions rob the excited atoms of energy before they have a chance to emit a photon of light. Greisen, with the help of Bunner and a group of graduate students, set out to put this knowledge to good use. They built the very first Fly's Eye detector in 1967.

The Cornell scientists built a multi-coloured, 25-sided building that looked more like a children's playhouse than a serious astrophysical observatory! Sixteen of the sides housed 500-millimetre diameter windows which were actually lenses for collecting and focusing the fluorescent light from the air showers. Any resemblance to a playhouse was forgotten when the mass of cables and electronics inside the building was seen. A smaller 16-sided structure inside the observatory held a total of 505 photomultiplier tubes, which peered through the lenses to collect the faint flashes of light. The system, similar in concept to the optics of a fly's eye, was arranged so that each photomultiplier tube viewed a particular part of the night sky. Electronic signals from the 505 photomultipliers were displayed on a bank of 505 oscilloscope screens which was photographed when a set of electronic circuits detected a possible cosmic ray signal in the photomultiplier tubes. Each 'pixel's' view of the sky was very conveniently displayed in this way. Students spent long hours peering at hundreds of rolls of these photographic records, in search of a convincing event.

Greisen and his students were looking for a signature of an air shower. This would consist of a light signal starting in one of the photomultipliers looking high in the sky and progressing down through a series of other pixels. With the technology of the day this was a difficult task, and Greisen drew on experience from the new field of high energy physics research to try to detect the faint showers. The task has been compared to detecting a 5-watt blue light bulb hurtling through the atmosphere at the speed of light! Unfortunately, Greisen and his team never saw an event they felt confident about. They were foiled by the lack of cheap, fast electronics and the lack of large, cheap light collectors. The idea was not dead, however. A young graduate student at Cornell, George Cassiday, was working at the time on a separate experiment, a more traditional high energy physics accelerator experiment. But he was excited about the possibility of identifying the great cosmic accelerator and he took a great deal of interest in Greisen's successes and ultimate failure.

## The Fly's Eye—the Utah version

Cassiday moved to Salt Lake City in 1970 to take up a post-doctoral position at the University of Utah. Apart from the intellectual activity he enjoyed in the cosmic ray group led by Jack Keuffel, Cassiday was able to pursue his lifelong interest in a whole range of outdoor activities, from running and hiking through to skiing and whitewater river-running. Even today, Cassiday is often seen on the streets of Salt Lake City running in a half-marathon. His cheeky sense of humour kept him sane in a city and state heavily influenced by the Mormon church. His response to door-knocking

Mormon missionaries was to offer them a beer—by that action he clearly displayed his interest in their visit, and they generally got the message!

The Keuffel group's forte was researching the interactions of high energy muons from air showers. Their large experiment was housed in a silver mine near the historic town of Park City, now more famous as the site of the yearly Sundance film festival run by Robert Redford. The group was interested in how high energy muons interacted with the rock above the detectors, not so much in the astrophysics of the primary cosmic rays. They were taking advantage of the particle beam from the cosmic accelerator, like so many of the earlier pioneers of high energy particle physics. But in 1972, Cassiday realised that it was possible to combine the study of astrophysics and particle physics with a detector like the Fly's Eye. He convinced Keuffel that Greisen's idea needed re-examination and jumped into a year-long design study. A series of calculations and computer simulations told Cassiday that, with more modern high-speed electronics, the idea was now feasible. In 1975, the group received government funding and trials began. Tragically, Jack Keuffel died that same year from a heart attack while climbing in the mountains. Cassiday lost his greatest mentor but was able to assume the leadership of the group to take the Fly's Eye project to a successful conclusion.

Cassiday realised that apart from better electronics, the new Fly's Eye needed to have larger light collectors. The 0.5-metre diameter lenses of Greisen's detector were replaced by 1.5 metre curved mirrors, which focussed the light from the night sky onto a cluster of photomultiplier tubes. However, the character of the detector was the same—each photomultiplier viewed a unique element (or pixel) on the sky. Because the government funding agency wanted proof that the new concept would do better than its predecessor, Cassiday and his team took a small number of mirrors and prototype electronics down to Albuquerque, New Mexico, to the site of John Linsley's Volcano Ranch array. During the clear New Mexico nights, they pointed their detectors over the array and waited for large air showers to arrive. After several false starts, the experimenters were delighted to detect fluorescent light at the same time that Linsley detected shower particles in his scintillators. Not only were coincident measurements a success, but Cassiday and Linsley agreed that the two techniques determined the same ground-level shower size, to an accuracy better than 10 per cent.

These encouraging results guaranteed funding for the entire project. It was time to find a permanent home for the Fly's Eye. About 140 kilometres west of Salt Lake City, just next to the Skull Valley Indian reservation, lies the giant US Army Dugway Proving Grounds. Established for more than fifty years, and covering a significant fraction of the north-west corner of the state of Utah, Dugway offered a number of advantages to Cassiday's team. For one thing, there was electric power and other infrastructure in an otherwise isolated area of desert, far away from city lights that would swamp the sensitive photomultiplier tubes. The presence of army security would mean that vandalism would be no problem. The only negative with the location was the occasional military exercises that restricted the access of scientists to the site for a few days every year. In 1977, the construction of the full 'eye' began on top of Little Granite Mountain, a 130-metre-high hill that offers a superb view of the surrounding desert and more distant mountains. The Utah team constructed 67 mirror units, each with a cluster of either 12 or 14 photomultiplier tubes in the mirror focal plane. The mirrors were pointed in such a way that each of the total of 880 photomultiplier tubes viewed a different 5-degree-diameter hexagonal section of the night sky. The mirrors were housed at the ends of lengths of 2.1 metre-diameter corrugated-iron pipe, rather like water tanks. At the beginning of a night of operations, motors rotated the housings so that each mirror pointed towards its designated portion of the sky. During the day, the mirrors were directed towards the ground, protecting them and the photomultipliers from the weather and the fierce sunlight.

**Figure 8.2  The Fly's Eye image of a cosmic ray shower**

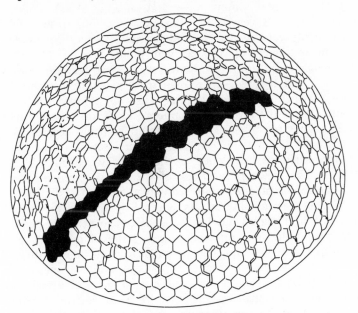

The original Utah Fly's Eye detector collected images of the night sky with 67 mirrors, each mirror containing a cluster of photomultiplier tubes. In the diagram, each hexagonal pixel represents the view of a single photomultiplier. An air shower is detected when a series of photomultiplier tubes sees a light signal, as shown by the filled pixels.

Like experiments that detect Cerenkov light from air showers, the Fly's Eye had a limited viewing time per month, with operation only possible on clear, moonless nights. With perfect weather, this would mean that operations would be possible for 18 per cent of the hours in any year. In practice however, the Fly's Eye operated roughly 12 per cent of the time—but this reduction in operating time is compensated for by the enormous area coverage of the technique. The Fly's Eye was able to view a volume of atmosphere extending out 20 kilometres from Little Granite Mountain—so an area of 1000 square kilometres was viewed 12 per cent of the time, making its sensitivity equivalent to a ground array of over 120 square kilometres operating 100 per cent of the time.

Inside the Fly's Eye control room, just near the computers and twenty 2-metre-high racks of electronics, stood a blue hemisphere of plastic. This object attracted the attention of the Fly's Eye operators and visitors alike during the nightly operation of the detector. The hemisphere, about one and a half metres in diameter and set into the wall, represented the hemisphere of the night sky above the Fly's Eye. The viewing directions of the 880 photomultipliers were represented by 880 red lights set into the plastic. Whenever an air shower was detected by the Eye, the event would be replayed on the hemisphere, known as the 'light display'. The operator would see a series of red lights go on in succession, showing the passage of the shower down through the atmosphere. For this display, the speed of passage of the shower was slowed by a factor of a million, to allow the inadequate human eye to see the event! Meanwhile, the computer was recording information from all the photomultiplier tubes that viewed the shower, including the intensity of light viewed by the tube and the time of arrival of the light at the tube, to an accuracy of fifty-billionths of a second.

Information on the pointing directions of the tubes that 'fired' and the time of the firing was later used to determine the position of the shower axis, the imaginary line in space along which the shower developed. Knowing the axis and the amount of light seen at each position along the axis, it was possible to calculate how many particles were present in the shower at all stages of its development. The unique ability of the Fly's Eye to directly measure this so-called 'development curve' was its major strength and allowed the Utah scientists to estimate the energy of the original cosmic ray in a very straightforward way. The development curve also gave them an excellent handle on the particle's mass. During each night's run, the operator could stare at the light display and be assured of the detector's smooth operation by the beautiful showers that would flash up every minute or so. He or she was sometimes surprised to see words spelled out on the display that was meant for showers! It was George Cassiday who occasionally programmed the computer running the light display to randomly flash up messages that would always amuse, especially those operators used to Cassiday's earthy humour!

While the Fly's Eye found it impossible to see the faint fluorescent light from air showers at distances of more than 20 kilometres, the detector saw other, much more distant, sources of light. One night during the construction period, Cassiday turned on a new bank of mirrors that happened to point towards the northern horizon. To his horror, he found that several hundred photomultipliers in the detector regularly fired at intervals of 1.5 seconds. Cassiday and his colleagues went outside to look in the direction of the supposed light, but couldn't see a thing. A couple of nights of detective work followed, and the mirror positions were moved to try to pinpoint the direction more exactly. It was then a matter of jumping into a car and driving towards the supposed source. Eventually, the scientists found an aircraft-warning strobe light on top of a smokestack at the International Lead Company smelter on the shore of the Great Salt Lake, 100 kilometres from the Fly's Eye! The company was not in a position to switch off the light, so Cassiday was forced to design a circuit that made the Eye 'blind' to any triggers for a fraction of a second every 1.5 seconds. (Eventually, the smelter stopped operations and the smokestack was demolished when the bottom fell out of the lead market.)

A more spectacular source of background light was discovered in a more public fashion. A television crew from the Salt Lake City CBS station had set up their camera in the Fly's Eye control room, and was sending live pictures back to the station during the ten o'clock news program. The science reporter was describing the function of the Fly's Eye light display when suddenly half the red lights on the display lit up, indicating a huge flash of light in the sky. The University of Utah scientist hosting the camera crew's visit, a normally quiet and restrained graduate student named Bob Cady, let out an exclamation not often heard on the local TV news. All rushed outside to see the sky lit up by what turned out to be a failed Soviet satellite re-entering the atmosphere in a sensational fashion.

Someone from the Salt Lake City branch of the Federal Bureau of Investigation must have been watching Channel 5 that night. The FBI called the university the next day to ask for a representative of the experiment to call at their offices. They had imagined a sinister use of the Fly's Eye, a use that the Utah scientists had not given much thought to—the monitoring of the sky for the re-entry of intercontinental ballistic missiles. The FBI was worried that the 'other side' might have seen the advantages of the Fly's Eye. Cassiday selected Peter Gerhardy to visit the FBI office. The agents interviewed the Fly's Eye representative, quizzing him about any approaches the group had received from foreign nationals. Gerhardy, in his broad Australian accent, replied that he had been in contact with a whole heap of foreigners lately, but that they had all been American!

The Fly's Eye operated successfully from its construction in 1978 through to 1993. Halfway through that period, the Utah scientists added a second eye to the system. The so-called Fly's Eye II was constructed a little over 3 kilometres from Little Granite Mountain on the valley floor, and consisted of 36 mirror units covering only half of the night sky—that half in the direction of the original eye. It was always on the books that the group would build a second site, but it was a matter of waiting for the funding from the National Science Foundation. In 1967, Greisen had seen the advantage of having a stereoscopic view of each shower and Cassiday's group was in the position to implement the idea. As we've seen, the first step in calculating the energy of a cosmic ray is to determine the line in the atmosphere, the axis, along which the shower develops. With a single Fly's Eye, this is done by first noting which photomultipliers in the eye saw the light from the shower. It's convenient to think of the firing tubes as they are displayed on the hemisphere of the light display. These tubes form a line on the night-sky hemisphere, and this defines a plane in space—that plane containing the shower axis and the Fly's Eye. The plane is called the shower-detector plane. The traditional Fly's Eye method then uses the times at which each of the tubes fired to mathematically determine the orientation of the shower axis within the plane. This determines the shower path geometry.

If, instead, two separated eyes view the same shower, then each eye can define its own shower-detector plane. In a much quicker analysis procedure, the shower axis is simply the line in space defined by the intersection of the two planes. A stereo view of showers produces a more accurate position for the shower axis, because it doesn't rely on the tube firing times that can be affected by problems like imperfections in the mirror. An improved shower axis estimate, together with two views of the shower development curve, leads to a better energy assignment for the primary cosmic ray. From 1985 onwards, the two detectors viewed many showers in stereo over the western Utah desert, giving Cassiday and his team a large set of beautifully measured showers. However, because of the slightly special geometry required for a shower to be seen this way, there were many other showers seen only by the original eye, Fly's Eye I. These 'stereo' and 'mono' data sets were used for different purposes in the quest for more information on the mysterious cosmic ray beam.

### The Akeno giant air shower array

Before we discuss what the Fly's Eye and these other large installations have discovered, we need to describe the latest in the line of operational giant arrays. In 1975, a group of Japanese universities started work on the first of a series of ever larger experiments based in the farming area of Akeno, 200 kilometres west of Tokyo. As the array has grown, from 1 square kilometre to 20 square kilometres in 1984

141

and to 100 square kilometres in 1991, the scope and the importance of the observatory has also grown. It is now the largest ground array ever built and one of the largest scientific experiments the world has seen. Based on the traditional ground array, the 100-square-kilometre AGASA (Akeno Giant Air Shower Array) employs over 100 plastic scintillator detectors for measuring the air showers as they hit the ground. An additional set of 30 concrete-covered detectors is designed to measure the penetrating muon component of the showers.

The experimenters have been blessed and have also suffered in the semi-rural Akeno area. On the one hand, the placement of detectors in the middle of pastures and in the backyards of town dwellings has required great skills in negotiation, skills not required by Linsley at Volcano Ranch or Cassiday at Dugway. On the other hand, the presence of infrastructure in the form of a network of roads and power lines has made the powering and servicing of the detectors easy. It was even possible to connect each detector to the central data collection site with optical fibre. The AGASA detector is now a state-of-the-art instrument, a ground array with unsurpassed sensitivity and versatility. It has made some important discoveries and still has another five to ten years of useful life. We will describe some of these discoveries below. The central questions are clear—what are these cosmic rays, and where do they come from?

## What do we know about the highest energy cosmic rays?

After thirty years of measurements, a remarkable situation remains. There is no strongly preferred arrival direction for the highest energy particles. Linsley was the first to see hints of this result at Volcano Ranch, but it took the next couple of decades of hard work by many groups for this result to become more certain. The early expectation that the plane of the Milky Way would light up like a beacon as a source of these particles has not transpired. There are as many cosmic rays coming from the direction of the

Milky Way as any other direction in the sky. In other words, the arrival directions appear to be very isotropic. The interesting thing about this situation is that it doesn't make sense! If cosmic rays *were* accelerated to enormous energies within our own galaxy, then we would expect to see more events from the direction of the Milky Way. This assumes that cosmic rays move in more or less straight lines at the highest energies, with the galactic magnetic field too weak to affect their paths. It's possible that we've got the strength of the magnetic field wrong or our assumption about the charge on the majority of the cosmic rays wrong, but it's unlikely, as we'll see. On the other hand, perhaps the particles come from outside our galaxy. We expect there to be even weaker magnetic fields in the space between galaxies, so particle paths should be quite straight even over very large distances. So would we then expect to see arrival directions from all parts of the sky with equal probability? The answer is probably not.

It's certainly true that if you take a powerful optical telescope and point it at any area of the sky, you are bound to see a galaxy in your field of view. It might be a normal galaxy like our own or it might be a powerful 'radio' galaxy or quasar. These objects are fairly uniformly distributed over the sky. If cosmic rays were produced in a fraction of these objects, wouldn't we see events from every direction in space? We would, except for the ever-present and fundamental microwave background, the Big Bang remnant. Recall that this radiation places a limit on how far a highly energetic cosmic ray can travel through space, and in fact it places a very strong limit. We estimate that a distance of 300 million light-years is about as far as a particle could travel, though most would not get nearly so far. Most of the galaxies that we view through our telescope are more distant than this, and the close-by galaxies are very clumpy in their arrangement in the sky. The fact that we don't see a similarly clumpy cosmic ray arrival-direction distribution seems to put paid to a simple extragalactic origin model. It appears that neither of the simple-minded models, of galactic and extragalactic origin, can tell the whole story.

143

It's entirely possible that the true picture is a mixture of the two and, indeed, the results from the experiments have given support to such a mixed origin theory.

We saw earlier that one of the basic measurements that the experiments have made over the years is the cosmic ray energy spectrum. The spectrum is a measure of the relative number of cosmic rays that exist at each energy, and we've seen how rapidly the spectrum falls as we go up in energy. As we move up in energy by a factor of 10, the number of cosmic rays above that energy falls by a factor of about 100. This rule of thumb is remarkably accurate over an enormous range of energies, ranging from the copious low energy cosmic rays through to cosmic rays with ten million times more energy. Another way of expressing this is to say that the spectrum is a remarkably featureless graph, but it's the slight departures from the smooth spectrum that tell us so much. In the range of energies measured by the giant arrays and the Fly's Eye, there's a spectrum feature at around $3 \times 10^{18}$eV that marks a change in the slope of the spectrum. Above this energy the spectrum falls less rapidly and the feature is known as the 'ankle' of the spectrum. All the experiments have measured this feature, but the best picture of the situation came from results published by the Fly's Eye group in 1993, showing detailed data collected in stereo mode.

The Fly's Eye results offered something exciting and unique. The clear picture of the ankle was accompanied by results on the cosmic ray mass through this energy range. With the Fly's Eye's view of the entire shower development, it is possible to classify the mass of the primary ray as being light (like a proton or a nucleus of helium), medium (something like a carbon or oxygen nucleus) or heavy (for instance, an iron nucleus). The classification is based on the depth in the atmosphere at which the shower reaches its maximum size. As we've mentioned before, showers initiated by heavy primary particles develop quite rapidly and reach their maximum size at a shallower depth than proton showers. This behaviour is easiest to understand if you consider an iron nucleus

144

as a collection of 56 tightly bound protons and neutrons. When an iron cosmic ray with energy $E$ enters the atmosphere and collides with an air molecule, those 56 protons and neutrons are immediately released in the fireball of the collision. Each of these 'nucleons' carries off an energy of roughly $\frac{E}{56}$, and each initiates a sub-shower. Each sub-shower, only having a small fraction of the energy of the original, develops quite rapidly and the total air shower is the sum of these sub-showers. In contrast, a proton with energy $E$ will create a single shower that takes more time to reach its maximum size. The stereo Fly's Eye classified showers in terms of their 'depth of maximum size' and the Utah scientists found a fascinating correlation with the onset of the ankle in the energy spectrum.

The results were published in the prestigious scientific journal *Physical Review Letters* in 1993, and attracted the attention of many in astrophysics. The Utah group had found that the typical mass of cosmic rays changes as we move from lower energy particles at around $10^{18}$eV to the higher energies at $10^{19}$eV. It appears that there is a significant percentage of heavy particles, like iron nuclei, at the lower energies, with a gradual change to a light composition, mostly protons, at the highest energies. This change in the cosmic ray mass happens in the same energy region as the spectral ankle, leading to much speculation about its meaning. The Utah group put forward a particularly simple explanation. The idea is that the low energy part of the cosmic ray beam has its origin within our own galaxy and the high energy part comes from outside our galaxy. Given that two types of sources are likely for the two different components, it seems likely that each would have a different energy spectrum shape. The steep, low energy spectrum simply adds to the flatter extragalactic spectrum to give the observed composite graph with the characteristic ankle. The preponderance of protons in the high energy part of the beam is seen as reasonable in the model, since protons are better able to survive the passage through large distances in intergalactic space. Heavy cosmic rays like iron may be produced in those distant galaxies, but they would be readily broken up into pieces (ultimately separate protons and neutrons) through collisions with the photons of starlight that fill all of space.

This simple theory, which gives some explanation for the spectral and mass characteristics of the highest energy cosmic rays, also makes some sense when we think about the uniformity of their arrival directions. With this picture in mind we would not expect the Milky Way plane to stand out as a preferred arrival direction. Although most of the lower energy heavy particles are produced in the galaxy, their paths through the galactic magnetic field are particularly tortuous. (An iron nucleus, with a charge twenty-six times larger than a proton, has a very complicated path through the messy galactic field and it could easily appear to arrive at Earth from a direction completely opposite to its source direction, even at energies well above $10^{18}$eV.) At the highest energies, the proton cosmic rays don't originate in the Milky Way and therefore don't appear to arrive from those directions. The model assumes that there is a uniform distribution of extragalactic sources in the sky that would produce a uniform distribution of arrival directions at the highest energies. It doesn't specify the nature of these sources or the nature of the galactic sources at lower energies. Yet, as a starting point, this simple picture is extremely appealing because it fits the existing experimental results so well.

### Giant galaxies and tiny spinning stars

This model of cosmic ray origin assumes galactic and extragalactic sources without being specific—for a reason! Nobody has yet come up with a convincing way to explain how a star, a pulsar, a black hole, or an entire galaxy can accelerate particles up to energies seen in the highest energy cosmic rays. We know the sorts of things that are necessary in these sources, but no candidate sources appear to satisfy all the requirements. While this might sound like an appalling indictment on the cosmic ray community, it's seen by many as the most enticing aspect of the chase! Cosmic

146

ray astrophysicists really believe they are on the verge of uncovering something totally unexpected, something exciting. The Nobel prize-winning physicist Richard Feynman made the following comment in 1973:

> . . . when everything is so neatly wrapped up . . . with all experiments in exact agreement with each other and with the theory . . . one is learning absolutely nothing! On the other hand, when experiments are in hopeless conflict—or when the observations do not make sense according to conventional ideas, or when none of the models seem to work . . . one is really making progress and a breakthrough is just around the corner!

Feynman was talking about laboratory-based particle physics, but his message certainly has resonance with the situation concerning the highest energy cosmic rays.

Acceleration of charged particles requires either magnetic fields or electric fields. The electric field might be caused by a separation of electric charges in an astronomical environment. If, for some reason, all the positive charges in a volume are separated from all the negative charges, a potential difference (or voltage) will be created and a strong electric field will form. A proton released from the positive charge cloud will be attracted towards the negative cloud by a strong force and the particle will accelerate and gain a large amount of energy. However, if no electric fields are present, acceleration is possible, in principle, with magnetic fields.

As we've already seen, magnetic fields can affect the paths of moving charged particles. If the magnetic field is strong enough, it can actually act like a mirror and cause the particle to completely change direction. One old and well-respected idea in cosmic ray acceleration, first put forward by the particle physics pioneer Enrico Fermi, seems to apply in regions of space containing moving magnetic fields. If a cosmic ray particle is reflected off an approaching magnetic field region, it will gain some energy from the moving region and will be accelerated. Only a tiny amount of energy is accumulated by the particle on every reflection so the acceleration process is excruciatingly slow. To gain a significant amount of energy, a cosmic ray must bounce around in a region of moving magnetic field bubbles for millions of years. It might appear that, given enough time, a cosmic ray could acquire any amount of energy from this Fermi process. Unfortunately, there's a catch. In every situation, there will come a time when the particle becomes so energetic that it will no longer bounce off the magnetic mirrors. Instead, the particle will slice through the magnetic barriers, ending the acceleration process. Of course, this depends on the strength and dimensions of the magnetic field. If the field fills a large enough volume, even high energy particles will eventually be bent around and may continue their acceleration.

There's a good theoretical indicator of whether a particular astrophysical environment will be a likely site of cosmic ray acceleration under this Fermi process. It's the mathematical product of the average magnetic field strength and the overall size of the region. In other words, if B is the typical strength of the magnetic field and D is the diameter of the region containing the magnetic field, then there is a minimum value of B × D that will allow acceleration of particles to a particular energy. A large enough product can be achieved by having a large B, or a large D, or both. In all those cases the cosmic ray will stay long enough in the region for acceleration to that energy to take place.

The first thing to be said about this B × D rule is that it's an absolute minimum requirement for the Fermi acceleration process. Without a minimum value of the product, the situation is hopeless. In addition to a good value of B × D, we need the magnetic fields to be moving in some way, and the faster the better. A collision with a fast-moving magnetic field bubble will transfer more energy to the cosmic ray. What about possible acceleration sites of cosmic rays we see with energies of $10^{20}$eV? Remember, these particles are known to exist and they must be accelerated somewhere. The fact is that it is virtually impossible to find a place in the Universe that has the necessary BD product to get protons accelerated to this energy—even assuming that the magnetic fields are moving at close to the speed of light! This is the fundamental problem with the theory at the moment. There are two types of astronomical object that lie within the realms of possibility, although both are a real stretch! One is a type of object that we find in our Milky Way galaxy, the other is more exotic and distant.

Pulsars are examples of objects that possess very large magnetic fields, up to a trillion times larger than the magnetic field around our Sun. While they are incredibly tiny objects (only 30 kilometres across), the product BD is still large enough to squeeze them into consideration for $10^{20}$eV cosmic rays. The acceleration mechanism is certainly not clear. It would need to be very quick before the cosmic ray escaped the tiny environment. However, this scenario is at least conceivable in the presence of such large magnetic (and electric) fields!

The only other reasonable candidate source couldn't be more different to a pulsar—it consists of radio hotspots in the jets of active galaxies. These regions have a very weak magnetic field but their BD product is saved by the enormous size of the sites which are at least the size of normal galaxies. Here, the acceleration process would be a much more leisurely affair. The huge volume bottles up the cosmic ray for millions of years as it makes the occasional collision with a fast-moving magnetic irregularity. There is evidence that large amounts of material are ejected with great speed in the jets of active galaxies. The ejected material would carry magnetic fields along with it, a necessary ingredient for Fermi's mechanism. A radio hotspot is likely to be some sort of magnetic knot or irregularity in the jet. The hotspots appear to be sites of great activity, where it's certain that electrons are being accelerated to energies of millions of electron volts. The intense radio-wave activity is a result of the energetic electrons radiating energy as they spiral around the tangled magnetic field lines in the knot. Certainly a powerful region, but is it powerful enough to accelerate protons to energies a hundred trillion times higher than the electron energy?

The answer lies in the details—the speed of the magnetic irregularities, the uniformity of the general magnetic field, and the size of the region. It seems that these sites are the leading contenders for the highest energy cosmic rays. Both radio hotspots and pulsars are theoretically possible, but theorists are leaning towards the former because they allow more latitude in the values of the important parameters. We know less about radio hotspots than we do about pulsars, so it's easier to stretch the numbers to make them sources! How unlucky we are that there are no regions of space with both the size of radio hotspots and the magnetic field strength of pulsars. These regions would be able to accelerate particles to energies well in excess of those observed! Or do these regions exist somewhere? The history of astrophysics tells us that it's dangerous to say 'never'!

## The mysterious Cygnus X–3

While pulsars seem to be out of favour at the highest energies, they could be a better bet at $10^{18}$eV. It seems that we might have some experimental evidence in this area, uncovered by the Fly's Eye experiment and the AGASA array. It relates to a Milky Way X-ray star called Cygnus X–3. This object was the third X-ray source found in the constellation of Cygnus, discovered by rocket observations in 1966. Subsequent observations in 1970 by *Uhuru,* a NASA X-ray satellite (named with the Swahili word for freedom), showed that the X-ray signal from this object, like some other X-ray sources, varied in a periodic way. The periodic behaviour suggested that Cygnus X–3 was not a simple star, but two stars in orbit around each other. Astronomers believe that one of the stars is a neutron star that, with its strong gravitational pull, strips material from the atmosphere of its companion, a main sequence star. The stolen gas apparently collects in a disk surrounding the compact star. Friction heats up the gas, producing X-rays. As the large companion star orbits around the neutron star every 4.8 hours, it periodically blocks our view of the X-ray source. As this happens, the X-ray intensity waxes and wanes. Binary X-ray sources, like Cygnus X–3, are rare objects, with only a dozen or so known in the Milky Way and our neighbouring galaxies, the Small and Large Magellanic Clouds. As a class, they are some of the best-studied objects in high energy astrophysics, with emissions ranging in energy from radio waves to X-rays and beyond.

Cygnus X–3 is one of a few of these sources that has been observed emitting gamma-rays, but unfortunately the certainty of these observations is not clear because of the very low signal levels involved. However, if the gamma-ray observations *are* true, they would help solve some of the many mysteries in cosmic ray production. As we learned in Chapter 7, ground array observations of Cygnus X–3 in the early 1980s seemed to indicate that the source was emitting $10^{15}$eV gamma-rays. The most promising models of gamma-ray production in X-ray binary systems suggest that the gamma-rays are actually by-products of cosmic ray acceleration. Cosmic rays are produced in the classic environment of strong magnetic and electric fields, and gamma-rays are released as the newly energised particles exit the system. The cosmic rays produce mini-showers when they collide with gas nuclei in the companion's atmosphere, with gamma-rays being part of those cascades. In 1984, Michael Hillas of Leeds University looked at the gamma-ray signals from Cygnus X–3. Working backwards from these numbers, he showed that Cygnus X–3 *alone* could be responsible for accelerating *all* of the observed cosmic rays in the galaxy! Well, all of them up to energies of $10^{16}$eV, at least. This was an amazing thought and one that had people scrambling to get money to build larger and larger ground arrays to study Cygnus X–3 with much greater sensitivity. Unfortunately, as you'll recall, as soon as these arrays were completed, Cygnus X–3 turned off! There is still some debate about whether the source was *ever* on, but we know that the source is sporadic in other regions of the electromagnetic spectrum and many believe that it is also sporadic in gamma-rays.

151

It was during the heady days of Cygnus X–3 hysteria, in 1986, that Jerry Elbert and Paul Sommers, two members of the Fly's Eye group, were looking at the arrival directions of the showers seen by the detector. In their analysis they plotted the arrival directions of all the Fly's Eye cosmic rays on a map of the celestial sphere. They ended up with a colourful density map that showed which parts of the sky were producing most of the cosmic rays. Most of the variation across the map was caused by what are called exposure effects. There are some parts of the celestial sphere that pass over Utah every day, but there are some that aren't accessible by the Fly's Eye detector. For example, the centre of the Milky Way can only be viewed from the southern hemisphere. After Elbert and Sommers made corrections for the exposure effects, they found that the brightest point on the map, corresponding to the largest density of cosmic ray arrival directions, lay in the direction of Cygnus X–3!

Did this mean that cosmic rays were being emitted by the object? A careful analysis showed that the signal *could* have been a chance clustering of otherwise randomly directed cosmic rays, particles from other places in the galaxy. However, the probability of this happening was small, about 1 chance in 1500. This level of certainty was not sufficient to convince everyone that Cygnus X–3 was a very energetic source. On the other hand, there were many who believed that it was remarkable that the hottest spot in the sky just happened to coincide with one of the most studied sources in high energy astrophysics. The Fly's Eye data studied by Elbert and Sommers covered the period from the early 1980s through to 1987. It was an obvious step for other groups to check their data to search for a similar signal.

What transpired was an unsatisfactory, and puzzling, result. The AGASA group in Japan studied data from the various stages of the development of the Akeno array. They uncovered a signal from Cygnus X–3 that was of similar strength and statistical certainty as the Utah result. This was great news for the Utah experimenters and the spirits

152

of Cygnus X–3 believers were momentarily buoyed. Only momentarily. Very soon the news came from Alan Watson and the Haverah Park team that they had searched their data over the same period searched by the Fly's Eye and Akeno groups. They weren't able to find any signal from the source! The Haverah Park sky map was quite flat in that part of the sky, with no evidence of any excess of events from Cygnus X–3. The jury is still out on this strange inconsistency. If not for the Akeno result, the Haverah Park data would have convinced many that the Utah observation had been a statistical fluke and that the 1 in 1500 chance clustering had indeed happened. But the Akeno confirmation completely changes that argument and has kept people scratching their heads to the present day.

Assuming that the Cygnus X–3 result was real, what sort of particles were coming from the source? The signal was present at the lowest energies accessible by the Fly's Eye and Akeno arrays, at around $3\times10^{17}$eV. This is much too low an energy for protons to move through the galactic magnetic field in straight lines. The obvious first conclusion was that extremely energetic gamma-rays were initiating these air showers, making these the most energetic photons of light ever observed. But there was another possibility raised years before by several people, including Larry Jones of the University of Michigan.

Jones pointed out that neutrons, being uncharged, would move through magnetic fields unhindered. The only problem with free neutrons (that is, those not bound up inside atomic nuclei) is that they decay. In fact, if you place a neutron in a laboratory test tube, you will typically observe it decay into a proton and an electron within 15 minutes! On the face of it, this lifetime is not very promising for long-distance interstellar travel. But Jones and others pointed out that if the neutron is moving with a velocity close to the speed of light, time dilation in Einstein's Special Theory of Relativity would make it possible for free neutrons to travel across the galaxy! For a particle at Fly's Eye energies, the neutron's clock would appear to be ticking so slowly that it would take 30 000 Earth years for the clock to advance 15 minutes. In other words, Einstein's theory predicts that a neutron of this energy could travel 30 000 light-years through space without decaying. The signal from Cygnus X–3, a star system a little closer than 30 000 light-years, could conceivably be neutrons, and the neutrons would probably be by-products of charged particle acceleration in the source. Unfortunately, an air shower produced by a neutron looks much like one produced by a gamma-ray. It so happens, for example, that the atmospheric depth at which the two types of shower reach their maximum size is very similar at $3\times10^{17}$eV. So the true identity of the signal, if a signal existed, remains a mystery. Future experiments are planning ways of distinguishing neutron showers from gamma-ray showers. We look forward to their data on Cygnus X–3.

One last word on this intriguing object. The Fly's Eye has not seen any signal from the star system since 1987. On the other hand, the Akeno group has seen two episodes of emission in the early 1990s, both of which have been accompanied by very strong outbursts of radio waves. Cygnus X–3 has been known to be highly variable in the radio spectrum since September 1972, when a radio telescope in Ontario observed the first outburst. The binary star went from a very modest radio source to one of the brightest in the sky, with the radio intensity increasing by an incredible 1000 times in just a few days. Over the last two decades a handful of other radio bursts have been seen. During two of these, the AGASA array detected apparent cosmic ray signals. Unfortunately, neither of the signals have been very strong, nor were they observed by the Fly's Eye or Haverah Park. Haverah Park had closed down before these episodes and the Fly's Eye was unable to view Cygnus X–3 during these periods. The outbursts came during parts of the year when the source was in the same celestial hemisphere as the Sun and obviously not visible to the nocturnal Fly's Eye detector.

The activity of Cygnus X–3 inspired Australian physicists to check the records of the SUGAR array for signals from similar southern hemisphere sources. The SUGAR array, having been built in New South Wales, could not see the Cygnus constellation but a number of other X-ray binary star systems were visible. Unfortunately, the ability of SUGAR to assign directions to incoming cosmic rays was not as good as that of the northern experiments. The Sydney experimenters had opted for a sparse array to cover the largest area possible. However, a search was done for two objects, one a strangely named binary star in the constellation Corona Australis called 2A1822–37.1. The other was LMC X–4, a similar source in our neighbouring galaxy, the Large Magellanic Cloud. Both objects had apparently produced gamma-rays at lower energies and were seen as likely candidates. Each had a well-known period for the orbit of the two stars about their centre of mass (roughly 6 hours for 1822 and 1.4 days for LMC X–4). Remarkably, it was found that the SUGAR data arriving from the directions of 1822 and LMC X–4 were also modulated by the respective periods. That is, at certain times during the orbital period the array was more likely to record a cosmic ray. Again, the statistical weights of these results were quite low and they remain on the hairy edge of believability. But together with the Cygnus X–3 results, they point to an important class of objects that might be responsible for accelerating many of the cosmic rays in our galaxy, and an important target for the new detectors coming on line.

### The supergalactic plane—our intergalactic neighbourhood

Let's return to the cosmic rays with energies above $10^{19}$eV. As we've seen, the change in slope of the energy spectrum at around this energy creates the so-called 'ankle', and points to a different sort of origin for these particles. With these energies, and the small charge that we believe these particles have, they should rush through cosmic magnetic fields with very little deflection. Unhappily, until very recently, the arrival directions of these particles seemed very uniform in the sky—there was no significant hint of clustering of the arrival directions in any particular part of the sky. This was a great disappointment for everyone concerned. After all, most experimenters in the field, beginning with Linsley at Volcano Ranch and Wilson at Haverah Park, had expected that the origin of cosmic rays would be solved by their experiments within a decade or so. Thirty years on we are still confronted with a puzzle, though we have been able to rule out many scenarios. You would be hard pressed to find a cosmic ray physicist who believed that the highest energy particles are produced within the Milky Way. However, the fact that no significant clustering of arrival directions was found doesn't necessarily mean that sources lie in every direction. The problem is, of course, a lack of data. After about thirty years, only about 1000 cosmic rays with energies above $10^{19}$eV have been observed worldwide. Only 100 of these have energies above $6\times10^{19}$eV. Perhaps it's not surprising that no obvious source region stands out on a sky map showing the arrival directions of these particles. Statistical techniques are needed to see whether there are any subtle clusterings of arrival directions.

But where to look? If we assume that cosmic rays are produced outside our galaxy, but within other galaxies, then it would make sense to see whether cosmic ray arrival directions match with galaxy directions. If you go outside and look into the night sky, you'll see the beautiful Milky Way as a band of stars stretching from one horizon to the other. Outside that band, many of the stars that you see are actually distant galaxies. If you used a large telescope to plot out the directions of all galaxies, you'd find a pretty uniform coverage in the sky. In other words, the Universe is isotropic on a large scale. This is certainly not true on smaller scales since we know that galaxies tend to exist in groups. Our galaxy is a member of a group of twenty or so galaxies, imaginatively known as the Local Group! This group is dominated by our galaxy and the Andromeda Galaxy. Most of the other galaxies within the group are dwarf and irregular galaxies like the Magellanic Clouds, but they do form a distinct group about 4 million light-years in diameter. Not too far away, at a distance of about 50 million light-years, is the Virgo Cluster, a very rich collection of 1000 galaxies contained within a diameter of about 7 million light-years. Being relatively close, and so large, this cluster is a significant feature of the night sky in the northern hemisphere. The Virgo Cluster contains some very impressive objects, including three giant elliptical galaxies, each one of them large enough to swallow our entire Local Group!

You may not be surprised to learn that clusters of galaxies tend to exist in associations. For example, our Local Group and the Virgo Cluster are members of a cluster of galaxy clusters, known as a supercluster, with a diameter of about 100 million light-years. In fact, throughout the Universe we see a distribution of galaxies that reminds many of a handful of soapsuds. Galaxies form structures that resemble walls separated from other walls by volumes known as voids—those areas less populated by galaxies. So what does all this mean for cosmic ray origins?

The complicated nature of the structure of the Universe can be simplified for us, since we really don't need to look very far out. As we've seen, due to the Greisen–Zatsepin effect, the presence of the cosmic microwave background puts a limit on how far a cosmic ray proton with energy above $6 \times 10^{19}$ eV can travel. Being very generous, we might expect that the maximum distance would be 300 million light-years. Similar distance limits exist for cosmic rays that might be heavy nuclei instead of protons. Collisions with the same photons in intergalactic space will break them up. If we now limit ourselves to all galaxies closer than 300 million light-years, we find a relatively uniform spread but with some indication of walls and voids. In particular, there is one wall-like structure, discovered by the French–American astronomer Gerard de Vaucouleurs in the 1950s, that contains our Local Group, the Virgo Cluster and other nearby clusters. This wall is known as the Supergalactic Plane. Cosmic ray physicists were not particularly excited about this plane until the early 1990s when Peter Shaver, an Australian radio astronomer working in Europe, discov-

157

ered something quite significant. When Shaver plotted a sky map showing the positions of nearby strong radio galaxies (rather than all galaxies) he found that the Super-galactic Plane became even more obvious. In fact, it seems that all strong radio galaxies within 300 light-years lie within 20 degrees or so of that plane! Given that strong radio galaxies are prime cosmic ray source candidates, parts of the cosmic ray community began to sit up and listen.

In particular, Alan Watson and Jeremy Lloyd-Evans from the Haverah Park array and two creative theoreticians, Todor Stanev and Peter Biermann, decided to re-examine the cosmic ray data from Haverah Park, Volcano Ranch and Yakutsk. They were keen to see whether there was any hint of cosmic ray clustering around the Supergalactic Plane. The work, published in 1995, has raised a lot of interest in the astrophysics community since their examination of the data showed that there was indeed some clustering in Supergalactic directions. The question asked by the team was the following—if we assume that cosmic rays at the highest energies are just as likely to come from any direction in the sky, then what is the probability that the observed clustering has happened by chance? The assumption of uniform arrival directions is a convenient hypothesis to test and if this probability is small, it gives weight to the alternate hypothesis that cosmic ray arrival directions are *not* uniform. The calculations showed that the 'chance' probability of the observed deviation from uniformity was 0.035, a small number but not small enough to be *certain* that the Supergalactic Plane is a source of cosmic rays.

A doubt still lingers in the minds of many people that Watson and his colleagues have been unlucky and have just observed a random clustering of otherwise uniform arrival directions! Still, the work has encouraged others to treat the Supergalactic Plane as a possible source region.

Lisa Kewley, a student in Adelaide, together with her two supervisors, has looked to see whether there is any similar Supergalactic clustering visible from the southern hemisphere. The SUGAR array, being the only southern giant array ever built, had a view of quite different areas of the sky. It was able to see regions of the Supergalactic Plane not accessible to the northern arrays. Unfortunately, when Kewley looked at the arrival directions of the highest energy SUGAR events, there was no sign of clustering around the plane. Obviously this has raised some interesting questions. Are the northern hemisphere results wrong, or is the population of cosmic ray producing galaxies in the southern regions of the Supergalactic Plane different to that in the northern regions? Unfortunately, this impasse will have to remain until more of the highest energy particles are observed. Luckily, we won't have to wait another thirty years. Major new detectors are either being built or planned. These will significantly quicken the pace of progress.

## The high resolution Fly's Eye—'HiRes'

As successful as the Fly's Eye detector was, it didn't solve the fundamental problem of the origin of the highest energy cosmic rays. This was a great disappointment to many, not least to the driving force behind the experiment, George Cassiday. Having put in a huge effort over a large portion of his professional life, Cassiday's interests began to move in other directions in the late 1980s. He bought a fast red sports car and began writing textbooks and working in other fields of physics, including biomechanics. Everyone recognised that this was a great loss to the field, but they also saw that George was revelling in his change in direction. There were other senior members in the cosmic ray group ready to take up the mantle. These were Gene Loh, a former colleague of Cassiday's at Cornell University, and Pierre Sokolsky, a relative newcomer to the cosmic ray field who came to Utah after work in neutrino physics at Brookhaven National Laboratory in New York. It was at this time, in 1987, that plans began in earnest for the next generation Fly's Eye detector. Cassiday and other members of the group had been informally working on ideas for the new project for several years before this, but it was in that year that serious calculations began. The aim was to present a proposal to the funding agency, the US National Science

Foundation. The direction was clear—the new detector had to see showers at larger distances than the old one, so that the event rate could be increased.

The original Fly's Eye could detect the highest energy air showers over an area of 1000 square kilometres—this was not enough! The new detector would need to see an area at least five times bigger. An obvious way to see further was to increase the area of the light collectors—the mirrors—so that the weaker light from more distant showers could be captured. The new mirrors would be 2 metres in diameter instead of the 1.5 metre diameter used in the Fly's Eye, increasing the mirror area by a little under a factor of 2. What about the light-detecting photomultiplier tubes that sit in the focal planes of the mirrors? In the original Fly's Eye these phototubes had provided images of the sky in a series of hexagonal pixels, each about 5 degrees across. This coarse view of the sky needed improvement for a couple of reasons. First, the new detector was to see showers at very large distances where the length of the shower would seem quite small in angular terms. Given that most of the action in shower development happens in the bottom 10 kilometres of the atmosphere, a shower seen 30 kilometres away might only appear 20 degrees long. The entire shower would all fit within only four Fly's Eye photomultiplier pixels and so only four measurements would be taken of the brightness of the light and its arrival time. Second, even for closer showers, an advantage was seen in getting a finer view of the shower development. It would improve the ability of the detector in determining the arrival direction, as well as in calculating the energy and the likely cosmic ray mass. The Utah group decided that a pixel diameter of 1 degree was ideal from a scientific point of view and could be justified financially. However, when the pixel diameter is decreased by a factor of 5, the number of pixels required to cover the same area of sky increases by a factor of 25! In this way, the High Resolution Fly's Eye project, or 'HiRes', was born.

One thing was clear from the start of the HiRes planning—the instrument would need to be a 'stereo' device to continue the success started by Fly's Eye I and Fly's Eye II. As we've already said, viewing a shower with two separate Fly's Eye detectors significantly improves the measurement accuracy of the cosmic ray arrival direction and energy. The HiRes plan was ambitious. The group would build three HiRes sites on the corners of a triangle measuring 15 kilometres each side. This was much larger than the 3-kilometre separation of the old detectors and it indicated the increased seeing power of HiRes. It also expressed the desire of the Utah experimenters to concentrate on the highest energy (and brightest) cosmic ray showers. A site would contain a total of 78 mirrors, each with a focal plane cluster of 256 photomultiplier tubes. In other words, the HiRes design called for a phenomenal 60 000 photomultiplier tubes! This was not a project the Utah group could cope with alone. Pierre Sokolsky convinced his old colleagues in neutrino physics to join the project as collaborators. These groups from Columbia University and the University of Illinois, experienced in large experiments at particle accelerators, were able to bring engineering and physics input to the project. In 1992, the cosmic ray group from the University of Adelaide also officially joined the effort, after a long association with the Fly's Eye that had seen four Adelaide PhD graduates taking up research positions on the project during the previous decade.

The early 1990s has not been a good time to ask for big money from the US funding agencies. The price tag for the total HiRes project, around $15 million, has so far been elusive. However, the support from the prime agency, the National Science Foundation, has been as strong as for any other project on their books. Currently, the HiRes project has full funding for a scaled-down version of the project, consisting of a total of 72 mirror units at the two detector sites. This is known as Stage 1 and construction will be completed by 1999. (Stage 2 will complete the original vision and is likely to be funded at the end of construction of Stage 1.) One of the two sites is Little Granite Mountain, the site of the original Fly's Eye at

Dugway Proving Grounds. The second site is on Camel's Back Mountain, 13 kilometres away on the other side of a desert valley used by the US Army for munitions testing. Prototype mirror units have operated at the two sites since 1992 and the quality of the shower data already collected is surpassing all expectations. The HiRes collaboration members are looking forward to the day when Stage 1 is complete. At that time the collecting area of the detector will be in excess of 5000 square kilometres, with every one of the high energy showers seen by both HiRes sites. These will be the best measured cosmic ray showers ever. The data rate of 300 cosmic rays per year above $10^{19}$eV will put the group in the best ever position to finally sort out the issues around the ankle in the energy spectrum. Does the Greisen–Zatsepin cut-off really exist? Are the cosmic rays all protons at the highest energies? Do the arrival directions point back towards the Supergalactic Plane?

## The highest energy particle ever observed

The funding outlook for the HiRes project took a great boost early in 1993. At that time a young Chinese scientist named Dai Hongyue was in his fourth year with the cosmic ray group at Utah. Dai was analysing data from the original Fly's Eye in order to put together an energy spectrum. As part of the checking process, he was using his computer screen to display pictures of the air shower development in the atmosphere. For each event, the picture showed measurements of how the shower size varied as the shower passed deeper and deeper into the atmosphere. Dai was flicking through these plots quite quickly, keeping track of poorly analysed events. This was the first pass through the analysis for these showers so it wasn't surprising that some needed more work. Very often though, the events looked good, with measurements that produced beautifully shaped and well-proportioned profiles. Dai paused on one of these events and took a breather, feeling very pleased with how nicely this particular event had come out. His eye wandered to the vertical axis of the plot which showed the shower

**Figure 8.3 The most energetic air shower ever recorded**

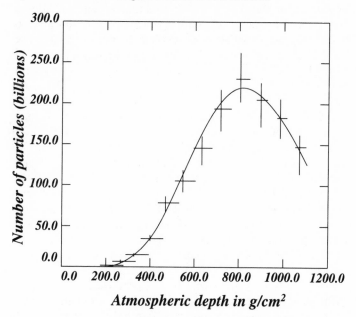

*Atmospheric depth in g/cm²*

The cosmic ray that initiated this shower had an energy of
$3\times10^{20}$ eV and was detected by the Fly's Eye. At its maximum size
this shower contained an incredible 200 billion particles. The
figure shows how the size of the shower changed as it travelled
down through the atmosphere.

size, the number of charged particles in the cascade. He
was amazed by what he saw. The shower size peaked at
an astounding 200 billion particles, a thousand times larger
than most showers seen by the detector. This was the
largest shower Dai had ever seen!

Dai called over Jerry Elbert and Paul Sommers from
the offices next-door and they began work on this event.
They first tried to find out what was wrong with the analysis
of the event! What mistake could make the shower look so
large? This was not a trivial process and it took several

weeks of work as the team investigated every possibility, however remote. Finally they concluded that the shower was extremely energetic, coming from a primary cosmic ray with energy of $3 \times 10^{20}$eV! All involved in the Fly's Eye were over the moon about this single event—the beautifully formed profile represented the highest energy cosmic ray (indeed, the highest energy elementary particle) ever observed!

While the development profile gave a very good estimate of the energy of the cosmic ray, it was more difficult to identify the mass of the particle. This is because, as we've seen, showers can develop in slightly different ways even if they are initiated by particles with the same energy and the same mass. The Fly's Eye depth of shower-maximum technique makes it possible to look at results from a *group* of showers. It's possible then to decide whether the group is better described as a sample of proton showers or iron nucleus showers or a mixture. However, it's much more difficult to say a similar thing for a single shower. The best guess for Dai's event is that it was produced by a nucleus with middling mass (maybe something like an oxygen nucleus), but this does not rule out it being a proton. A proton would be more in line with the expectation from the trends in the lower energy Fly's Eye data that we discussed earlier.

The particle had arrived on the night of 15 October 1991, and because of the complex analysis required for Fly's Eye data it remained buried until Dai uncovered it a little more than a year later. Its discovery was significant in many ways. It was far more energetic than the previous record-holder in the Fly's Eye catalogue. That cosmic ray, with an energy of $8 \times 10^{19}$eV, had been detected back in 1984 and it looked as though it was going to hold the energy record for ever. The Fly's Eye, unlike other detectors, had never previously detected energies closer to $10^{20}$eV and there was a feeling in the group that the Fly's Eye might prove the existence of the Greisen–Zatsepin cut-off. This was a feeling shared by many in the wider cosmic ray community. It was true that several showers had been reported by other groups with energies around $10^{20}$eV, including Linsley's Volcano Ranch shower in the 1960s and some showers from Haverah Park and SUGAR. However, there was some concern that the calibration of these experiments was a little uncertain, especially at these enormous energies. On the other hand, the Fly's Eye method, with its view of the entire shower development, was seen to be a better way of measuring energy. However, the entire argument was tipped on its head with Dai's event of October 1991.

The Fly's Eye energy spectrum now looks very strange with the graph showing a big gap between the two highest energy events. There is still a possibility that the Greisen–Zatsepin cut-off exists, with the microwave background radiation preventing most of the higher energy particles reaching the Earth. The existence of just one event, way above the cut-off energy, suggests that the source of that event may be 'local'. The particle didn't travel far enough through extragalactic space to lose much energy. It could be that some cosmic ray sources are closer than others. But why haven't we detected more of the super-high energy particles?

Soon after the announcement of Dai's huge shower, the Japanese group at AGASA discovered their own. In 1992, a particularly large shower had landed on the experiment, conveniently in an area of the array well covered by particle detectors. Like the Fly's Eye event, the AGASA shower was a beautifully measured one and it was relatively straightforward to determine an energy of around $2 \times 10^{20}$eV. So while the Fly's Eye group still held the record, there was added evidence that there were extremely energetic sources in our local intergalactic neighbourhood! You might think it strange that the two highest energy cosmic rays ever seen were discovered so close together in time. Well, we agree that it is a little surprising. On the other hand, the 100 square kilometre AGASA has the largest collecting area of any experiment currently in operation and had only been in operation for a year or two. So if any group was going to detect a rare, high energy event with a ground array it was going to be the Japanese team.

At the time of the Fly's Eye and AGASA announcements, there was enormous interest in the arrival directions of the showers. Wouldn't it be easy to trace back the paths of these particles in order to discover the sources? Well, yes and no. You'll recall that the bending of the cosmic ray path depends on the charge of the particle, the strength of the extragalactic magnetic field and the length of the path. Based on the best estimates of the strength of the field, the path of a $3\times10^{20}$eV proton would only be deflected by at most 10 degrees over a path of 150 million light-years. In other words, if the particle was a proton and if it had travelled the maximum possible distance through the microwave background, there wouldn't be much area of sky to search for a strong radio galaxy or some other plausible source. The Fly's Eye event appeared to come from a direction in the constellation of Auriga, almost opposite to the direction of the center of our galaxy. Unfortunately, there is no powerful radio galaxy within the obvious search region. There are two unremarkable galaxies closer than 150 million light-years and within 10 degrees of the arrival direction. They don't exhibit the jets and radio lobes that we suspect to be necessary for acceleration to this energy. On the other hand, if the search region is relaxed a little and one looks within 12 degrees of the arrival direction, one finds a strong radio galaxy. The object is called 3C134. Unfortunately, we can't readily estimate the distance to this galaxy. Our view of it in visible light is obscured by a giant gas cloud in our own Milky Way, making it difficult to make a spectral measurement to get the galaxy's redshift and hence its distance. Luckily though, the object is detectable in the radio spectrum (quite strongly), and the extent of the radio signal in the sky is large enough to indicate that the source is relatively close, quite possibly within 150 million light-years. There is now some pressure on the optical astronomers to make a redshift measurement of this galaxy, despite the technical challenges.

The AGASA giant shower was produced by a particle arriving from the constellation of Pisces. In this case we struck it lucky. There is a strong radio source, NGC 315, about 10 degrees from the measured cosmic ray arrival direction. Redshift measurements indicate that it lies 150 million light-years away, right on the upper distance limit. This galaxy is certainly a strong candidate for the source of the particle, simply because it's the only candidate!

With only two events in our catalogue of cosmic rays with energies well in excess of $10^{20}$eV, we have to admit that we can't be hugely confident that these particles are accelerated in the radio lobes of strong radio galaxies. This is despite the tantalising clues. It is quite possible that we are just being fooled by the proximity of the radio galaxy candidates to the cosmic ray arrival directions. This might just be a coincidence or perhaps we have got some of the assumptions wrong. For example, the paths of the cosmic rays may have bent more than we've assumed. The magnetic field out in intergalactic space might be stronger than we think or the cosmic ray particles in question might not be protons but more highly charged nuclei. All we can say at the moment is that we *think* that our estimate of the field is pretty right, and the overall Fly's Eye mass-composition results *indicate* that the highest energy particles are protons. The answer is surely that we need more observations of similar super-energetic particles to see whether the arrival directions cluster about the 3C134 and NGC 315!

## The X-particle

Since the publication of the details of the Fly's Eye and AGASA particles, this lack of absolute certainty about the radio galaxy origin has opened the floodgates on some alternative and interesting ideas. One of the most intriguing is not associated with traditional ideas of cosmic ray acceleration in astronomical objects like galaxies or pulsars. We label the traditional processes as 'bottom-up', where particles start with little energy that builds up slowly. Instead, the alternative theory says that super-energetic particles might appear out of nowhere! This is an example of a 'top-down' theory where a particle is created with its enormous energy already intact, through the decay of a

167

super-massive parent particle. In other words, the energy of the cosmic ray would come from part of the mass of the decaying particle. Such a process is, of course, allowable because of Einstein's equivalence of mass and energy. However, the decaying elementary particle would need to be extremely massive. An electron has a mass equivalent to only 511 keV and a proton has an energy equivalent about two thousand times larger, or about $10^9$eV. Our hypothetical 'X'-particle would need to be at least a trillion times more massive to produce the Fly's Eye particle.

It turns out that some so-called Grand Unified Theories, a class of theories that attempt to unify the fundamental forces of nature into a single description, predict particles with masses in the range of $10^{24}$eV! They would have been produced soon after the Big Bang and the standard theory would expect them to decay soon after production. However, some believe that a fraction of these X particles is trapped in 'folds' in the space–time of the Universe. These folds, called topological defects, would have some properties in common with those other notable defects in the fabric of space–time, the black holes.

The theory goes on to predict that a collapsing topological defect could release an X-particle at any time. The particle would then spontaneously decay, converting its mass into a number of particles with extreme energy. Those particles would include gamma-rays, neutrinos, protons and neutrons. Interestingly, a topological defect need not be associated with an astrophysical source like a radio galaxy or with any collection of normal material, for that matter. They could be distributed randomly throughout space and would produce energetic cosmic rays seemingly from nowhere. As such, the theory is a fascinating alternative to those that predict an association between cosmic ray arrival directions and clusters of galaxies. Given that a single X-particle is expected to decay into a swarm of energetic particles of different types, an experimental challenge for the future is to look for this signature in rare cosmic ray events. In fact, the expectation is that the majority of particles released in a decay would be super-energetic gamma-rays. A detector that could identify a gamma-ray initiated air shower would be a great plus. As it stands at the moment, because of the similarities between showers produced by protons and gamma-rays, it's entirely possible that the Fly's Eye and AGASA showers were initiated by examples of these wondrous photons!

One final note on the topological defect scenario. We've already indicated that there's a wide 'gap' in the energy spectrum measured by the Fly's Eye, between the event with an energy of $8 \times 10^{19}$ eV and the highest energy event. If this gap persists as new experiments collect more and more data, the case for the topological defect model would be strengthened. The model naturally predicts the production of super-energetic particles, but not particles of lower energy. The theory assumes that the lower energy particles we see are produced in the more traditional bottom-up acceleration processes. If, however, the bottom-up processes are responsible for the highest energy particles, we wouldn't expect to see any energy gap. After all, if we observed particles of $3 \times 10^{20}$ eV energies coming from nearby active galaxies, we would also expect to see particles with energies of $1 \times 10^{20}$ eV or $2 \times 10^{20}$ eV coming from the same objects.

# CHAPTER 9

# THE AUGER PROJECT
## —THE ULTIMATE COSMIC
## RAY OBSERVATORY

There is a trend occurring in many areas of science. Scientific projects are getting big! You can see evidence of the trend if you go into your local academic library and flick through a scientific journal. You're likely to very quickly see a paper with a huge list of authors. It used to be possible in many areas of science for one or two people to make great advances in their field. Think of Darwin and the evolution of the species or Hubble and Humason discovering the expansion of the Universe. While it still can happen this way, the advance of science has meant that in many areas experimental apparatus has become more complex and expensive. The number of people required to operate the experiment and analyse the data has increased. We see the most extreme example of this in high energy particle physics, where author lists on papers from experiments at accelerators like Fermilab in the US or CERN in Switzerland sometimes stretch to more than 400 people!

The trend towards larger and larger detectors has also been occurring in cosmic rays. From single experimenters like Hess and his balloon, through the small groups put together by Auger and Rossi, we are now moving through a period of rapid expansion. For example, the HiRes collaboration is now typical of the international collaborations operating in the field, with a team of about twenty-five astrophysicists. The next stage of growth is already being planned, to make its impact soon after the turn of the century. Called the Pierre Auger Project, this experiment will take over from HiRes and AGASA in the push to discover the origin of cosmic rays at higher and higher energies. The project will combine the efforts of over 100 astrophysicists from twenty countries.

The situation is this. Together, HiRes and AGASA will collect four or five hundred cosmic rays every year with energies between $10^{19}$eV and $10^{20}$eV. They will no doubt solve many of the outstanding problems in that energy range, including the cosmic ray spectrum (Is there a Greisen–Zatsepin cut-off at around $6 \times 10^{19}$eV?); the composition (Are the particles mostly protons?); and the origin (Do the particles come from the Supergalactic Plane?). But the recent observations of two super-energetic particles have opened up a whole new vista above $10^{20}$eV. Here the HiRes and AGASA outlook is not so rosy—they will be lucky to collect four or five of these particles every year between them. The Auger Project is planned to tackle this enormous challenge.

The leading lights in the project are Jim Cronin from Chicago, and Alan Watson from Leeds, two names that we've mentioned before. Cronin is the former high energy particle physicist and winner of the Nobel Prize in 1980 for his experimental work on a fundamental field of physics called Charge-Parity violation. He became interested in cosmic rays and high energy gamma-ray astronomy in 1985 and went on to build the CASA air shower array in Utah (see Chapter 7). Watson is the former leader of the now closed Haverah Park array. Together, they've gathered a worldwide collaboration of 100 physicists interested in building the Auger Project, estimated to cost $100 million. The physicists include leading contributors from the US, UK, Japan, France, Russia, Germany, Argentina and Australia. Some of the experimenters are joining the project from accelerator-based particle physics; others are currently working on a variety of the existing large air shower experiments. The fusion of the two groups of physicists looks like producing a very strong collaboration, easily capable of pulling off the many technical challenges.

The fundamental design criterion for the Auger Project

**Figure 9.1  The arrangement of detectors for the proposed Auger Project**

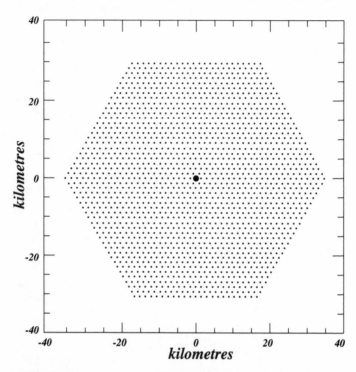

Each of the 1600 dots represents a single 10-square-metre area
detector separated from its neighbours by 1.5 kilometres. The large
dot at the centre represents a Fly's Eye-type fluorescence detector.

is a cosmic ray collection rate of 5000 showers per year with
energies above $10^{19}$eV! With our usual assumptions about the
way the energy spectrum might go at even higher energies,
this would lead to a very acceptable 50 cosmic rays a year
above $10^{20}$eV. Given the rarity of these particles, this is rather
a tall order but entirely possible! The plan calls for two
enormous arrays, each with an area of 3000 square
kilometres. In other words, each would be thirty times the
size of AGASA, the largest array built so far. One array will

be placed in the northern hemisphere and one in the southern hemisphere, so that the entire celestial sphere can be searched with close to uniform sensitivity. The southern site will be located in Argentina. That country has taken particular interest in the project and its President, Carlos Menem, has given his personal support. Very large areas of Argentina would be suitable, especially in the Patagonian regions where large flat areas are available with low rainfall and clear skies. In September 1996, the collaboration chose an area in Millard County, Utah (150 km south of the HiRes location), as the northern hemisphere site.

As we've seen, the experiments investigating the highest energy cosmic rays have been of two distinct types—involving ground arrays like Volcano Ranch, Haverah Park or AGASA, or using fluorescent light detectors like the Fly's Eye and HiRes. Each technique has its strengths and weaknesses. Fly's Eye detectors have the unique advantage that they can directly view the development of the air shower as it passes down through the atmosphere. This provides a very straightforward method for determining the cosmic ray energy and mass. However, Fly's Eye detectors have the unfortunate disadvantage of only being able to operate during dark nights when the weather is clear. The presence of a bright moon can even affect their ability to see the faint light from distant air showers.

On the other hand, a ground array can operate every minute of every day. Unfortunately, such an array has a limited view of air showers since it only measures the shower at a single, fixed level in the atmosphere—the ground. At that level, an array can measure each shower rather well, identifying the components of the cascade (muons, electrons and gamma-rays) and determining the way these particle numbers fall off as the distance from the shower core increases. This 'lateral' measurement of the structure of the shower is complementary to the 'longitudinal' measurement of the shower made by a fluorescent light detector. The Auger collaboration has seen the sense in combining the two types of detection methods in the same observatory. The Auger Project detector will

173

be a hybrid, combining the complementary strengths of a large ground array with a very sensitive fluorescent light detector.

The genesis of the Auger Project occurred at a regular meeting of the international cosmic ray community in Dublin in 1991. Jim Cronin planted the idea into people's minds at that time and over the next few years the project took form. Several workshops were held around the world during those years at locations that stressed the international form of the collaboration—Paris in 1992, Adelaide and Tokyo in 1993 and the Colorado mountain retreat of Snowmass in 1994. In 1995, a working group was assembled at Fermilab, the home of the major US particle accelerator. During a six-month design workshop, international scientists visited Fermilab to collaborate face to face, and at other times used the Internet to continue the work from their home countries. The result of this activity by 100 scientists was a very thick book known as the *Pierre Auger Project Design Report*. In great detail it discusses the proposed form of the project and gives justification for the ideas.

Each of the two world sites will have a ground array of 1600 detectors, each detector having an area of 10 square metres. The detectors will be like those used at Haverah Park, really just large tanks of water viewed by photomultiplier tubes that detect the passage of shower particles via the Cerenkov light they emit in the water. The choice of water Cerenkov detectors over detectors made from plastic scintillator material (like those used in AGASA, for example) was made chiefly on the basis of cost.

Both types of detectors, provided they are designed in the right way, have the ability to identify the different components of air showers at ground level. As we've seen, this is crucial for finding out the mass of the primary cosmic rays. The two critical portions of the shower are the muon component and the electromagnetic component— that component of the shower consisting of electrons, positrons and gamma-rays. In the case of the water Cerenkov detectors, the separation is done on the basis of the size of the light signals seen in the tank of water. It turns out that air shower muons have much more energy than their electron and gamma-ray companions. The muons are able to make it all the way through the tank of water (1.2 metres deep) without being stopped or absorbed. On the other hand, the electromagnetic component is absorbed in the top half-metre of the tank. Even gamma-rays result in a small amount of Cerenkov light. Soon after entering the tank they are likely to 'pair-produce' an electron and a positron, both of which will be quickly stopped in the water. Since the amount of Cerenkov light produced by an energetic charged particle is proportional to the length of its path in the water, muons can be identified by the bigger light signals they produce. Complex digitising electronics will be installed on each Auger detector to record these Cerenkov signals.

Each detector in the array will be very isolated. Its nearest neighbours will be 1.5 kilometres away. The grid arrangement of the detectors will resemble a set of inter-locking hexagons, with every detector surrounded by six others arranged on the corners of a hexagon. While this separation seems large, you'll recall that the showers at the highest energy are enormous. It's estimated that a $10^{19}$eV shower will typically hit a total of 6 detectors, and a shower ten times more energetic will produce signals in 18 detectors. Given the distances involved and the isolation of the likely Auger sites in Argentina, it's impractical and expensive to string up power and communication lines between every detector. Instead, the detectors are being designed to run on solar power and they will communicate with each other, and the array control centre, using high frequency radio. The Australian telecommunications company, Telstra, conceived a unique communications scheme especially for the Auger Project. It will use elements of mobile telephone technology to provide a very simple wireless network for collecting data and for checking the health of every detector in the array. After all, it won't be practical for a technician to visit every detector in the 3000-square-kilometre array more often than, say, every six months.

The fluorescence detector part of the Auger Project will probably be a single 'eye' at the centre of the array. This will depend a bit on the sites finally chosen for the two arrays. One eye will be sufficient if the topography of each site is such that 3000 square kilometres can be accommodated within a 62-kilometre-diameter circle (rather than a more elongated shape), and if the air is clear enough for good transmission of the weak fluorescent light. The eye will look somewhat like a HiRes site, though the mirrors will be four times larger than in HiRes in order to see well out to the edges of the array. If the topography or air clarity demands, two or three sites will be needed. In that case, the eyes won't be required to see as far and they can be built with smaller, cheaper mirrors. The cost of the multi-eye option won't be very different to the cost of the single, larger eye. The fluorescence part of the Auger Project is costed at about $20 million, or about 20 per cent of the overall cost.

It's possible that the fluorescence component of the project will be supplied by a single country, Japan. Professors Teshima and Nagano, two of the leaders of the AGASA array from the University of Tokyo, are currently working on prototypes of a new fluorescence project called the Telescope Array. Originally proposed as a stand-alone successor to HiRes, to be introduced at the turn of the century, it's quite likely that it will find a home as part of the Auger Project. This extremely ambitious design will be more than a replacement for HiRes. It will have large mirrors (3-metre diameter) with very finely segmented clusters of photomultipliers in every focal plane. The combination will allow it to see fluorescent light from very distant showers. However, they will have a second, quite different, use in the area of TeV gamma-ray astronomy, where the telescopes will look at showers with energies $10^8$ times smaller than those studied in the Auger Project! Each mirror will be steerable and during the higher energy observations each mirror will be pointed in a unique direction, producing the 'Fly's Eye' pattern on the sky. However, when TeV observations are planned, all the telescopes will point in one direction, for example towards the Crab Nebula, to detect the Cerenkov light from air showers initiated by $10^{11}$ or $10^{12}$eV gamma-rays. In this mode, the Telescope Array will be the most sensitive detector of its type with the ability to detect gamma-ray sources one thousand times dimmer than the Crab in just ten nights of observation. The only problem the experimenters will have will be to decide when to operate the telescopes in each of the two modes—both modes will be the biggest and best ever seen in their respective energy ranges!

Funding is being sought by members of the Auger Collaboration from governments around the world. Construction will commence at the beginning of 1998, with half of each array being completed two years later. At that time, the first significant data will begin to roll in, with the full operation of the arrays beginning in 2002. By that time HiRes and AGASA will be well towards completing their tasks and all eyes will be on the Auger Project for answers to the burning questions at the highest cosmic ray energies. There is no doubt that all three experiments will answer many questions. But experience in scientific enquiry has taught us something else—that those questions will be replaced by new ones, questions delving even deeper into the mysteries of the Cosmos.

# Appendix 1

# Some relativity

In the early years of the twentieth century, Albert Einstein recognised that our view of space and time was incomplete. He came to this conclusion through a consideration of the ways in which the laws of electricity and magnetism combined to produce the properties of electromagnetic radiation, such as light. He recognised that the speed of light must play a central role in physics if light was to have consistent properties in all measurements. In particular, the speed of light in a vacuum would have to be the same, 300 000 kilometres per second, no matter how the source of the light and the observer might be moving relative to each other. Einstein considered how things would appear when one moved at great speed. It would be our usual expectation that the speed of light would appear different depending on whether we moved in the same direction as a light wave or we moved in the opposite direction (or indeed at an intermediate angle). Remarkably, Einstein recognised that this could not be so, almost for purely aesthetic reasons.

In the seventeenth century, Newton had proposed a classical form of relativity when he argued that experiments (involving moving bodies) should not be affected by any background uniform motion of the reference frame in which they were carried out. Einstein saw that this seemed to clash with his ideas of electromagnetism and was perplexed when he tried to see how a light wave would look to an observer travelling at high speed. He saw then that, in order to get consistent answers in physics, one could not consider space to be just the volume in which we live. It also had to have properties which meant that as one travels at high speed, the scale of space would change while, at the same time, the scale of time changed. In this sense, space and time were intertwined and appeared to be related manifestations of the same thing, space–time.

We are well aware that, in normal life, we do not see this distortion of space and time. That is because we do not deal with speeds remotely like those of light. In fact, relativistic phenomena depend on the ratio of a speed to the speed of light *squared*. This ratio is only important when, say, one is considering speeds above a tenth or so of the speed of light since only then does the ratio rise above one-hundredth. This speed regime is almost limited only to the experience of the high energy physicist. Since we are not used to such speeds, the results of special relativity seem strange to us. In reality, they are certainly somewhat complex (although described by equations which are well within the reach of high school mathematics) but they have always been found to work perfectly and at low speeds they are almost exactly the same as the familiar rules of physics.

Physicists rarely have to use the full repertoire of relativistic ideas. Here, we will only discuss two—time dilation and the equivalence of mass and energy. Time dilation is a favourite example of a relativistic effect and was first observed with cosmic rays. We noted that in relativity the scales of space and time change with the speed of the observer. If, for instance, we make measurements of a clock which is moving towards us, we find that it runs more slowly than it would have if it had been stationary next to us. On the other hand, if we had travelled with the clock at its speed, it would have appeared to keep good time. We do not see ordinary clocks travelling towards us close to the speed of light, but radioactive decay is like a clock since it involves a well-defined time scale, the half-life. When we measure cosmic ray muons coming towards us, we find that their half-life is much longer than the 2.2

microseconds which we measure in the laboratory. In this sense, the internal clock of the muon runs more slowly from the point of view of ourselves as observers. Time is spread out—dilated.

Perhaps the most fundamental idea of physics is that of conservation of energy. If we allow a ball to roll down a hill, it speeds up and increases the 'energy' associated with its rolling and forward motions. This is its 'kinetic' energy or energy of movement. The increase in kinetic energy as the ball speeds up is exactly equal to the reduction in another property associated with the position of the ball in the gravity of the earth (as it travels to a lower height). The latter property is the 'potential' energy. If we add the kinetic energy and the potential energy, we get a fixed number—the total energy. Physics says that this total *never* changes; we just trade one form of energy for another. 'Energy' is thus conserved.

Over the years, this treasured conservation 'law' has had to be extended to incorporate some other phenomena, for instance heat. This is all pretty well understood and accepted. Now think what happens when a nucleus disintegrates. One second, the nucleus just sits there and the next second it flies apart. Suddenly, there is kinetic energy where there was none before. Something else has to be added to the energy total to account for the apparent increase in energy. A careful accounting shows that the mass of the disintegrated nucleus is less than the original and the energy discrepancy is always accounted for if we say that the extra energy needed equals the mass lost multiplied by the speed of light (usually written as 'c') squared. We say that mass and energy are equivalent with a constant of proportionality which equals $c^2$. That is, $E = mc^2$—perhaps the best-known formula in physics.

Einstein reached this conclusion theoretically by demanding that energy and momentum still had to be conserved as he changed the concepts of space and time to space–time. It was a great triumph of the theory when it was found that the formula worked exactly. In cosmic ray physics we are always dealing with high energy particles.

Their kinetic energies are always greater than the mass energy they would have if they were stationary (their rest mass). This means that their mass is always variable and very close in value to their kinetic energy divided by $c^2$. As a result, we tend to talk about the energy of particles and forget about the rest mass. For instance, we might talk about an electron with an energy of 50 MeV. The rest mass of an electron (usually expressed for convenience in energy units) is about 0.5 MeV and so, even at these modest energies, the rest mass is only about 1 per cent of the total mass/energy of the electron. We thus tend to consider only the total (or even just the kinetic) energy.

In the case of the 50 MeV electron, its energy in joules (the usual energy unit and the one required for use in Einstein's equation) would be 50 x $10^6$ x 1.6 x $10^{-19}$ joules. We can convert this to the total mass of the particle by dividing by $c^2$ i.e., 3 x $10^8$ x 3 x $10^8$ which gives us about 9x$10^{-29}$kg. This is tiny but is still about one hundred times the mass of an electron one would find quoted in a textbook.

The Einstein mass/energy relation is crucial in cosmic ray physics for two reasons which we can identify from the previous example. First, in principle, we can convert the 50 MeV of electron energy into the mass of other particles provided that there is a suitable physical mechanism. This is exactly what a cosmic ray shower does. Very high energy cosmic rays have their kinetic energy converted to mass in the form of multitudes of *real* particles so that a single primary particle can cause a shower of secondary particles. Second, the mass equivalent of the energy *really is mass*. For instance, when we calculate the path of a cosmic ray particle in a magnetic field, its spiral curve is of a size which requires the inclusion of energy divided by $c^2$ as mass in the calculation to give the correct answer. The appropriate value for the mass is the total mass *including* the energy. In cosmic ray studies, the energies are often so extreme that we do not even bother adding in the real 'rest' mass.

# Units and scales

## Distances

Cosmic ray studies deal with very large distances which are usually measured in the same units as other distances in astronomy. These units are either light-years or parsecs. A light-year is simply the distance travelled by light (in a vacuum, or, for practical purposes, in space) in one year. It is about $10^{16}$ metres. This is clearly a huge distance but it is still conveniently small for astronomical purposes. For instance, the Sun is about 8 light-minutes from us, the nearest star is a few light-years away and our own home galaxy is tens of thousands of light-years in diameter. A brief reflection on the true meaning of this in terms of the vastness of space quickly places our human experience in perspective. For historical reasons, professional astronomers use a related unit of distance, the parsec (pc), which has a value of a little over three light-years.

At large distances, such as those between galaxies, we have a fundamental uncertainty in our distance measurements since they are often based on the rate of expansion of our Universe. The exact value of this rate is the subject of great debate and is not agreed upon to much better than a factor of 2.

## Electron volts (eV)

Cosmic ray energies are measured in electron volts (eV). One electron volt is the energy gained by a single electron as it travels through one volt of potential. For instance, each electron moving through a circuit between the terminals of a dry cell will change its energy by about 1.5 electron volts. This value is tiny. A joule (the standard unit of energy) has a value of about $6 \times 10^{18}$ electron volts. The lowest energy cosmic rays which are measured near Earth have energies of about 1 billion electron volts (sometimes written as 1 GeV) and the highest energy cosmic rays known have energies above $10^{20}$ eV.

## Grams per square centimetre (g.cm$^{-2}$)

Cosmic rays are progressively absorbed as they pass through material and it is convenient to be able to say how much material has been crossed. This is done by imagining a cylinder with an area of 1 square centimetre around the track of the particle. We measure the matter passed through by the mass of material in this cylinder. Its units are grams per square centimetre (g.cm$^{-2}$). This rather strange way of measuring distance turns out to be very practical since one is usually interested in the number of times the particle has an interaction in its passage. This number is large for dense material and small for light material if the distance is measured in metres but, if it is measured in these units, the number of interactions is pretty well proportional to the number of g.cm$^{-2}$ travelled.

As a consequence, we can see that the atmosphere of the Earth, with a thickness of about 1000 g.cm$^{-2}$ from sea level, has very similar absorbing properties to a 10-metre depth of water. By the way, you might have guessed this fact since atmospheric pressure is the same as the pressure of 10 metres of water. If you dive in the ocean to a depth of 10 metres, you double the pressure on your body (atmospheric plus 10 metres of water) and also double the absorption of cosmic ray particles.

# OUR ATTEMPTS TO MATCH THE COSMIC ACCELERATORS

Great advances in our understanding of the structure of the atom were made during the first two decades of this century. Studies of the atomic nucleus began with Rutherford's famous 1908 experiment which showed that the atom was composed of a small, positive nucleus surrounded by a sea of electrons. Rutherford used high energy alpha particles (now known to be helium nuclei) from the radioactive decay of radium to probe the atomic structure of a thin sheet of gold. At that time, radioactive decay was the only source of high energy particles for this purpose. Of course, cosmic ray secondary particles were ever-present, and many of the early discoveries in particle physics were made with this cosmic debris. But the cosmic rays are an untidy collection of particles of varying mass, energy and direction. With a radioactive source like radium, it was possible to produce a particle beam of fixed energy and mass. By using shielding, it was also possible to collimate the alpha particles into a pencil beam. All these things helped in the early studies of the structure of the atom.

However, a typical radium source could not provide a high rate of alpha particles. Physicists began to think about machines capable of producing high energy particles. As the Nobel laureate Luis Alvarez explained of the period,

> the tedious nature of Rutherford's technique repelled most prospective nuclear physicists . . . One microampere of electrically accelerated light nuclei would be more valuable than

the world's total supply of radium—if the particles had energies in the neighbourhood of one million electron volts.

The problem was that at this time there was no known way of achieving such energies.

Physicists knew that a positively charged particle, like a proton or an alpha particle, could be accelerated in an electric field. Experiments had been performed with electrodes placed at each end of an evacuated glass tube. The two electrodes had been connected to the terminals of a large voltage source, say 10 000 volts. (The vacuum inside the glass tube was a necessity. If air had been present, sparks would develop between the electrodes, current would flow and the voltage source would be shorted out.) A proton introduced at one end of the tube would be attracted to the electrode with the lower (more negative) voltage. The attraction would accelerate the proton to an energy of 10 000eV in our example. Unfortunately, this was nowhere near the goal of a million electron volts. By the early 1930s, the highest voltage source feasible was in the order of 30 000 volts. And it was realised that there were other technical difficulties associated with the acceleration scheme we described. Even if a voltage source of 1 million volts could be found, it was unlikely that the vacuum inside the glass tube could be made perfect enough to stop sparks flying between the electrodes. In other words, the idea of simple 'one-shot' acceleration to a million electron volts was seen to be impractical.

It was the young American physicist, Ernest Lawrence, who stumbled on the solution late in 1929. Lawrence first realised that particles could be gradually accelerated by passing the particles between a series of pairs of electrodes, each pair with a moderate voltage between them. Almost immediately he asked why he couldn't just use two pairs of electrodes and somehow channel the particles past those two pairs again and again. A magnetic field was the answer. A charged particle with particular velocity will move in a circular path in a magnetic field of the appropriate orientation. So Lawrence designed a particle accelerator that contained two places where a high voltage would accelerate the particles and a magnetic field that kept the particles circulating across those voltages time after time. The 'Cyclotron' was born, and after two years of tuning, a machine one-third of a metre across produced 1 million electron volt protons in February 1932.

In the more than sixty years since the creation of Lawrence's first machine, accelerators have become much larger and capable of accelerating particles to huge energies. The largest still use the basic idea of the Cyclotron, but with one important difference. This was necessary because the size of a particle's orbit in a magnetic field is proportional to the particle's speed and inversely proportional to the magnetic field strength. In Lawrence's machine the strength of the magnetic field was fixed, which meant that as the particle was accelerated the size of the particle's orbit increased. In more modern machines, known as Synchrotron accelerators, the strength of the magnetic field is increased as the particle's energy increases, so that the path of the particle remains a circle with a constant radius. Particle accelerators now consist of an evacuated pipe, maybe only 5 centimetres in diameter, bent into a huge circle kilometres in circumference. Strong electric fields exist at several places around the ring to boost the energy of the orbiting particles. Magnets are also distributed around the ring to keep the particles orbiting within the pipe, with the field strength increasing as the particle's energy and speed increase.

One example of a modern Synchrotron machine is the Tevatron collider at the Fermi National Accelerator Laboratory (Fermilab) in the American state of Illinois. The ring has a circumference of 6.3 kilometres and is equipped with 1000 super-conducting magnets. These powerful magnets, cooled to $-270°$ C, produce the extremely strong fields necessary to keep relativistic particles moving in a circular path. About 400 000 orbits of the ring are necessary to raise the energy of a swarm of protons to the maximum energy of 1 TeV, or 1 million million electron volts.

The Tevatron accelerates about 10 billion protons at once and they travel around the ring in a tight bunch. At the same time, 10 billion antiprotons (particles identical to protons except that they have a single negative charge instead of a positive charge) are injected into the ring. Being oppositely charged, the antiprotons move in the opposite direction around the ring, being accelerated in the same electric fields as the protons. At the end of the acceleration process, a swarm of 1 TeV protons is allowed to collide with the swarm of 1 TeV antiprotons moving in the opposite direction. Huge detectors surround the interaction point to observe the new particles produced in each of the proton–antiproton collisions.

The largest accelerator currently operating is at the European Centre for Nuclear Research, CERN. The LEP collider is 26.7 kilometres in circumference and accelerates electrons and positrons to 50 GeV in a tunnel 100 metres below the French–Swiss border. Plans are currently under way to use the same tunnel to build a proton–antiproton collider, with a maximum energy of 7 TeV for each beam. This Large Hadron Collider, or LHC, will use the most advanced super-conducting magnet and accelerator technology ever employed and will be commissioned in 2004.

The LHC looks likely to be the largest man-made accelerator in the foreseeable future. The even larger accelerator proposed by US scientists, the Superconducting Super Collider (SSC), was to be over 83 kilometres in circumference and would accelerate protons and antiprotons to energies of 20 TeV. This US$10 billion project was cancelled by the US Congress in 1993.

While these Earth-bound accelerators are superb instruments and technically impressive, they pale into insignificance against their cosmic counterparts. The highest energy cosmic ray particle ever observed had an energy of $3 \times 10^{20}$ eV, one hundred million times more energetic than the highest energy particles currently being produced in the Tevatron. The likely nature of the source of this particle is discussed in Chapter 8.

# GLOSSARY

**accelerator**  A device designed to increase the speed and energy of a subatomic particle. Cosmic accelerators are responsible for the production of cosmic rays.

**active galaxy**  A galaxy emitting unusual amounts of electro-magnetic radiation, e.g. Seyfert galaxies and quasars.

**AGN**  Active galactic nucleus. The core of an active galaxy responsible for the enormous luminosity. May contain super-massive black holes.

**binary X-ray system**  A pair of stars in close mutual orbit. One star is a compact object (neutron star or black hole) which draws gas from its companion star into a hot accretion disk that emits X-rays.

**black hole**  An object with gravity so strong that the escape velocity is greater than the speed of light.

**Cerenkov light**  Light emitted by a medium (e.g. air, water) when traversed by a charged subatomic particle moving with a velocity greater than the speed of light in that medium. The light is emitted in the direction of motion of the inducing particle.

**cosmic microwave background**  A sea of radiation that bathes the entire Universe. A remnant of the Big Bang, the radiation currently has a characteristic temperature of 2.7K.

**cosmic rays**  Extremely energetic atomic nuclei (including protons) from space that bombard the Earth.

**electromagnetic cascade**  A cascade in the atmosphere consisting of electrons, positrons and gamma-rays and initiated by a high energy gamma-ray. An extensive air shower initiated by a cosmic ray particle will contain many electromagnetic cascades, each initiated by a gamma-ray from a pion decay.

**electromagnetic radiation**  The family of radiation including (in order of decreasing energy) gamma-rays, X-rays, ultraviolet light, visible light, infra-red radiation, microwaves and radio waves.

**energy spectrum**  A description of the range of energies present in a sample of cosmic rays, and the number of particles present with each energy.

**extensive air shower**  A cascade of subatomic particles in the Earth's atmosphere, initiated when an energetic cosmic ray collides with an atmospheric atom or molecule.

**flourescent light**  Light emitted through the excitation and ionization of atoms and molecules by high-speed charged subatomic particles. This light is emitted in all directions, unlike the beamed Cerenkov light.

**galaxy**  A large collection of stars, nebulae, interstellar gas and dust.

**gamma-rays**  The most energetic form of electromagnetic radiation.

**ionization**  The process by which an atom loses electrons. May occur when a charged high-speed subatomic particle passes through a collection of atoms.

**isotope**  Any of several forms of a chemical element whose nuclei have the same number of protons but different numbers of neutrons.

**muon**  A subatomic particle of the lepton family, a sister particle to the electron. Exists in positive and negative varieties, and has a mass of 200 times the electron mass.

**neutron**  A subatomic particle with zero charge and a mass very close to that of a proton.

**neutron star**  An extremely compact star composed almost entirely of neutrons at a density equal to that of an atomic nucleus.

**pion**  A subatomic particle that exists in three varieties (neutral, single positive charge or single negative charge) with a mass approximately 250 times greater than that of an electron.

**quasar**  An apparent star-like object which is actually an extremely active galaxy at a very large distance, i.e. a quasi-stellar object.

**redshift**  The Doppler shift of light or radiation from receding astronomical sources. If the redshift is the result of the expansion of the Universe, it can be used to estimate the source distance using Hubble's Law.

**secondary cosmic ray**  A high-speed subatomic particle that exists as part of an extensive air shower initiated by a 'primary' cosmic ray.

**spectrum**  The result of dispersing a beam of light into its component wavelengths.

**supernova**  An explosion at the end of the life of a massive star which may leave behind the stellar core in the form of a neutron star or black hole.

**white dwarf**  An exposed and inert stellar core remaining at the end of the life of a low-mass star. The star has a radius similar to the Earth's radius.

# BIBLIOGRAPHY

Fazio, G.G. & Silbeberg, R. (eds) 1993, *Currents in Astrophysics and Cosmology*, Cambridge University Press, Cambridge

Friedman, Herbert 1990, *The Astronomer's Universe*, Norton, New York

Kaufmann III, William J. 1994, *Universe*, W.H. Freeman, New York

Rossi, Bruno 1964, *Cosmic Rays*, McGraw Hill, New York

Sekido, Yataro & Elliot, Harry (eds) 1985, *Early History of Cosmic Ray Studies*, Astrophysics and Space Science Library vol. 118, D. Reidel, Dordrecht

Wilson, John G. 1976, *Cosmic Rays*, Wykeham Science Series, London

# INDEX